D0853485

OIL

A GUIDE THROUGH THE TOTAL ENERGY JUNGLE

OIL

A GUIDE THROUGH THE TOTAL ENERGY JUNGLE

PHILIP WINDSOR

GAMBIT • BOSTON

1976

OCEAN COUNTY COLLEGE
LEARNING RESOURCES CENTER
TOMS RIVER, N.J. 08753

338.27
W766o

Copyright © 1976 by Philip Windsor

All rights reserved including the right to reproduce this book or parts thereof in any form

LCC: 76-1557 ISBN: 0-87645-090-7

Printed in the United States of America

First published in Great Britain by Maurice Temple Smith Ltd. and Heinrich Hanau Publications Ltd.

CONTENTS

75268

PREFACE

OIL : A GUIDE

When it was first suggested that I might write this book, my reaction was to protest that I knew nothing about oil. On reflection, that seemed to be an excellent reason to write the book, especially for one who professes to teach international politics. The question of energy, the oil industry, the relations between the states which have oil and those which buy it, have become so central to the problems of international relations, that there is no excuse for going on knowing nothing about them.

I therefore decided to find out, and to write, as far as possible in the language that I myself might understand, something about the world's dependence on oil, the way the oil industry works, and the nature of the changes of the past few years, which are often wrongly called the "energy crisis". Finding out proved fairly easy. There are now a great many books on oil. Writing proved very difficult. It kept showing me what I had not yet properly understood. I hope that the result is intelligible.

I am obviously indebted to a great number of books and people. My debt in particular to Professor Peter Odell will be obvious to anyone who has read his *Oil and World Politics.* I should also like to mention especially an anonymous counter-report called *The Oil Fix,* which has some fascinating bits of information. My particular thanks are also due, in the section on alternative forms of energy, to an American report by Dr. Robert E. Hunter; and more generally to a lecture I heard recently by Mr. Robert Mabro of St. Antony's College, Oxford. It was then that I realised I might begin to understand the oil question.

That question is still changing. The discovery of new resources, the changes in political circumstance, the relations between oil and other forms of trade and aid between rich and poor countries will certainly mean that the whole problem continues to evolve. For example, it is possible that in the future China will be as important a factor in the politics of the world's oil as Saudi Arabia is today. Oil discoveries in China certainly make this a possibility. But their significance cannot yet be judged in terms of the relationship between energy, politics and economics which I have described as they seem to operate now. This is therefore a book for today.

At which point, of course, I have to admit that the responsibility for what follows is mine and nobody else's.

Philip Windsor
January 1976

PART ONE

IT WILL NEVER BE THE SAME AGAIN

1

HOW OIL IS USED

We have heard a lot about economic miracles in recent years. The Japanese miracle and the West German miracle are the two most familiar to anyone who reads the newspapers, but journalists have also discovered the Dutch miracle, the Italian miracle (for a while), the French miracle (very impressive for some years) and even the Spanish miracle. But in fact, there has only been one economic miracle: the gigantic boom which was enjoyed for a generation by all those industrialised countries of the Western world which have been bound together in a common trading and financial system. These countries – the United States and Canada, Japan and the countries of Western Europe – have been getting richer without interruption, and helping each other get richer and avoid any interruption, for about twenty-five years. They have produced more and more goods, they have increasingly traded these goods with each other, and by doing so they have produced more and more money to keep the system going. Different individual countries have joined in this miraculous process at different moments and with different degrees of effectiveness; but even those which, like Britain, have played only a modest part in it, have still enjoyed a prosperity which had never been dreamed of before. It has been the most spectacular boom in human history.

Until some four years ago, it seemed to many people as if the boom could go on for ever – provided only that the governments involved showed good sense enough to help each other out of their occasional difficulties, and made the necessary adjustments in the system of trade and money. This *was* a matter of good sense as well as good will, for recent history had shown that when countries scored economic successes off each other, when they protected their own interests at the expense of other people's, when, in short they played beggar-my-neighbour with their economics, *everyone* was liable to suffer in the not very long run. This lesson, the argument went, had been learned the hard way by the experience of the years before the Second World War, and so now a sensible co-operation could be taken for granted, and with it a continuing prosperity.

It was so generally taken for granted that much of the discussion among economists in the Western world was about how soon, or in what conditions, the prosperity could be extended to other countries – those of the poverty-stricken majority of the human race. When would *their* economies be sufficiently advanced for them to multiply their own wealth faster than the demands they made on it, (which, in the jargon, is called reaching 'take-off point') and how could *they* be integrated into the system which was working so well? There were some rather Victorian assumptions behind this picture of the world, the assumptions of rich Victorians looking at the lower classes. Aid to the poor countries might be a necessary measure but so was a healthy dose of self-help; and the general expectation was that the rich would stay rich, but the poor might be able to join them if they worked hard and behaved properly. Among the poor, of course, were the Arabs and other oil-producing countries, who had not failed to notice the attitude behind the analysis.

Even though the attitude and the analysis were those which generally prevailed, some doubts had been raised. Certain voices, notably in France, had suggested that the system of money, on which trade between the developed countries was based, was essentially a temporary arrangement, and could not survive any radical adjustments in the relations between different currencies. Other voices, though very few indeed, and coming mainly from the edges of the scientific community, argued that there were simply not enough natural resources in the world for everyone to go on getting richer, and in fact that the developed countries were already using up at an alarming rate the natural wealth which it had taken the Earth several million years to lay down. But on the whole, the first of these arguments was shrugged off as an example of French bad manners; and as for the second, — well, there were always scientists to suggest that the world had only just begun to tap its real resources, and that there would always be enough new discoveries to keep the whole thing going until the final dream was realised: infinite supplies of inexhaustible power from the hydrogen atoms in the ocean.

But the third kind of warning was also uttered from time to time. It was that while, *in the end,* there might be enough natural resources to keep the human race alive and well, there was bound to be a shortage of basic foodstuffs and other materials while the world's population continued to grow, and while the richer countries continued to consume so much so ravenously. This, it was argued, would lead to a situation where the competition for those materials among the more rich and powerful nations meant a tremendous increase in prices (a commodity inflation, in the jargon) and where ultimately no nation, however rich, could stand the strain. This in turn would threaten the collapse of the world's money system, and that would mean the end of the prosperity which everyone had come to take for granted. *This* argument has been far more difficult to dismiss over the past three to four years — because it has begun to look like coming true.

The difficulty with these varied arguments, and the

difficulty with trying to judge whether the system of prosperity that has been built up over the past generation can really endure, is that they have all become mixed up with each other. Is the world really running out of natural resources? Nobody really knows. But is it running out of the kind of resources which it can use at present? That is a different kind of argument, because the nature of the resources that can be used at present depends on the kind of technology that is available now, the kind of technology that will become available over the next few years, and the amount of money that can be used to switch from one to the other. What, then, about the money? Take a very highly advanced industrial country, with massive programmes of scientific and technological research going on in half a dozen fields, which might all be of the utmost consequence for the future of the world. It can afford to pay for these so long as the economy is doing fairly well and the programmes do not have to be completed in too much of a hurry. But suppose that in order to keep the economy going, it has to start buying basic materials at a much higher price than before. The money has to come from somewhere, and probably among the first things that will be cut in order to produce it will be the advanced research programmes. This might mean that, instead of being able to say: We can produce new sources of food, or new kinds of power, in six years' time, it will have to say: We can't produce any alternative to what we are using now for at least fifteen years. Or even: We don't know when we will be able to see any alternative. And the country might be driven to this position because all the others are in the same boat as itself – which means that everything they produce and sell to each other has become dearer, that there is correspondingly that much less real money available for other things, that the whole system by which they maintained their wealth is in danger and that they are therefore bound to rely more and more on those other countries (usually the poorer ones) which produce the basic materials they need. So to the uncertainties of fundamental research are added the uncertainties of economic performance, and along with them the certainty that everything will cost more. This in turn could mean that the whole system of co-operation which had worked until then might be so endangered that it would produce its own opposite: a world in which countries were trying to outsmart each other by doing the best deals they could with the poorer nations that had what they needed; but that in doing so, they would be most likely to bring about a general increase in prices once more, and so add to all the difficulties they faced to begin with. A few years ago, this would have looked like an economist's nightmare. It is now a realistic prospect.

It is a realistic prospect above all in the area of one basic material, whose price had fluctuated previously, but has risen dramatically since 1973. It is also a material which is so basic to the economies of the developed countries that no alternative is in sight for several years. And it is also the material which, by its nature and its manifold uses, brings together all the forms of argument outlined here. The world *could* run out of it; but that is not likely to happen for a long time yet. But meanwhile demand for it has been rising so fast that the world could run out of the more cheaply accessible reserves before too long. And even if not, the possibility that this might be so, and the realisation that meanwhile there is no alternative, have combined to wreck the system by which it was originally obtained cheaply. Now the rise in prices even threatens to wreck the whole money system of the Western world. Yet even so, the first response of many countries to this situation was to try to secure advantageous deals for themselves, or at most for their associates. There was no attempt to work out a new system for the world as a whole. And this in turn has only increased the economic strains and will probably ensure that the research programmes on which a usable alternative depends, are cut back. For ten or fifteen years at the least, there is therefore not going to be any alternative worth speaking of, and it is most unlikely that there will be a general alternative until the end of the century. The question that has to be faced, in consequence, is not so much one of how soon an alternative can be developed, but what is to be done over the minimum period of ten years, to ensure that the economic system survives.

The material is, of course, oil. All modern economies, and all the programmes for the development of the poorer countries, depend on it. Most of the rest of this chapter will be concerned with the manner on which countries have come to depend on it – to show why it is so basic. But it is worth remembering that the oil question which has captured the headlines in the world's press is important not only in itself, but also as a symptom of what has happened to the economic development of the world: of the staggering imbalance that has developed between the progress of the advanced countries and the poverty of the rest; of the fact that the rest have now realised that the rich nations now really depend on them, and are determined to make them pay for it; and of the question which the rich and advanced countries have been most reluctant to face, and have still not begun to answer – namely, can their economies still work when they no longer have cheap materials from the rest of the world to use up at will? *That* is why the oil question is so vital: it shows, in the most naked and dramatic way, that the economic miracle of the last generation was based on a vulnerable and fundamentally unstable set of relations between rich and poor, in which the poor gradually came to realise their own power, and to demand something in return for their materials. And that also explains why the problems of oil are so hard to solve: they are related to a whole complex of political and social questions, in which the assumptions of the rich and the expectations of the poor still barely allow any meeting of minds.

But oil is also the central question in this whole new set of developments because of its own intrinsic importance. Why has it become so important?

After all, neither the industrial revolution, nor the imperial conquest by the European powers of most of the rest of the world depended on oil. Oil was not even seriously considered as a form of power until about 1880, and by then the industrial revolution was already old and the imperial system, which the industrial development of the Western world made possible, was firmly established. But the basic fact is simple: all industrial progress consists in making machines, first, do the work of men and do it more efficiently; and, second, produce goods and do work that could not otherwise be made or done at all. This in turn frees men to do other things, whether to conquer those parts of the world where the majority are still working in the fields, or to design even more machines and produce even more wealth. In other words, the power to drive machines becomes the basic requirement, not only of the economy, but of the whole political and social system that goes with it. The original power was steam, produced by boiling water over coal. But steam had serious disadvantages. It was a cumbrous form of producing power, it tended to mean large and heavy machines, and it also meant that large numbers of men had to work underground digging up the coal. As gas and electricity came to supplement and in some ways replace steam as major forms of power, it became simpler to design more efficient machines, as well as forms of heat and light; but at first both gas and electricity were still produced from the use of coal – that is, from the labour of miners. Even with machines at *their* disposal, miners were unlikely to produce all the coal needed for the power that was in such growing demand. And equally, supplies of good coal were far from inexhaustible; or else more and more money had to be invested in sinking mines where the coal was of reasonable quality but hard to get at. Which meant, naturally, that power threatened to become more expensive. As power becomes more expensive, so in an industrial society does everything else – which again threatens an economic crisis. (The standard way of dealing with this before the Second World War was to make things cheaper again by paying people less money, either physically, or by reducing the number of people who were earning money, through economic programmes that implied a large measure of unemployment. Which is one reason why miners tended to feel the brunt of such policies first: if they got paid less, power remained relatively cheap. Equally, other people, with less money available, could not afford to spend too much, so prices came down again, and everyone got through until the next crisis.) But what was amply clear was that cheap power was the source of all economic welfare.

Now the cheapest form of power that has ever been discovered is oil. The history of oil between the 1880s and about 1970 is the history of a growing dependence, among all industrial societies, on oil at the expense of other forms of power. The basic facts are as simple as that. More and more countries have switched from dependence on other forms of power, usually coal, to dependence on oil, for a wider and wider range of uses. And if they had no oil of their own, they imported it from countries which had. And if the countries which had it showed little interest in getting it out of the ground, the importing countries got permission to do it themselves – usually on very profitable terms. The background to the present problems with oil is that the countries which produce and export it are now trying to change the terms on which it is done..

Oil is cheap in three ways. First, it can be put more

easily, and more efficiently, to a wider range of uses than any visible alternative. Second, it is comparatively easy to get out of the ground – and although it demands a large and skilled labour force to ensure that it keeps flowing, it does not require anything like the heavy and dangerous work underground that coal does. Compared with the number of miners, the labour costs and the human toll that a ton of coal represents, the equivalent amount of power in oil can be achieved for practically nothing. And third, it is relatively cheap to transport – since it flows. This implies of course that it can do much of its own transport, along immense pipelines, and even the oil which is transported in bulk by sea can be pumped ashore and distributed conveniently; whereas coal demands much more in terms of railway, shipping, rolling stock, motorways, and lorries to deliver the equivalent amount of power. In all these ways, oil is so much significantly cheaper than coal that any economy which relies in the end on cheap power to keep going is bound to use more and more oil. Generally speaking oil demands a great deal of money to extract and move at the beginning – the initial capital costs; but it can make enough money to cover these very rapidly, produce huge profits for oil companies, and *still* be the cheapest form of power there is.

So far, I have used the word 'power' to describe the function of coal or oil, and I meant in doing so to emphasise that it was the widespread dependence on machinery which has increased the demand for power, and that it is in this sense precisely the "advanced" nature of modern industrial countries which has made them so dependent on poorer and less developed ones. But to continue to use the word 'power' would be misleading, since oil in fact provides energy, of which power is only one form. Energy really implies the ability to release power, whether to move things – as in the power to drive machines – or to be transformed into heat and light, or even food. And the first of the three measures by which one can see how cheap oil is lies in the fact that it is the most readily transformable deposit of energy yet available. It is the manifold abilities of oil in so many different fields which make it so convenient and therefore cheap.

First, it can burn directly, which means it can heat houses and other buildings. So can coal; but the burning of oil also implies much more than that. Because of its fluid nature, it can be burned in a much more controlled manner than coal, being pumped or sprayed in regulated amounts through pipes or small openings. This is indeed the principle of the car, which is propelled by a controlled explosion of ignited petrol. It is a relatively efficient use of energy, and even more important, this use can be compressed into a small machine. A coal-driven vehicle is not impossible, but it would have to be the size of a bus; and when it comes to aircraft, only oil – or paraffin, derived from oil – can do the trick. Equally, coal has been largely replaced by oil for railway engines, which has taken the romance out of the railways, but made them cheaper and more efficient to operate.

Second, apart from burning directly, oil can produce other forms of energy through its heat; and oil-fired power stations for the generation of electricity are now nearly as common (in some countries more common) as coal-fired plants.This means that a very high proportion of heat and light, as well as of the machine-driving power, in advanced industrial countries is now derived from oil (Britain, for instance, uses 90% of its oil imports in this way). Together these two qualities of oil indeed imply that by far the greater part of a country's transport system, and a very large proportion of its industrial system now depend, directly or indirectly, on oil.

Beyond such obvious importance, though, oil has tremendous uses in other applications. The molecules of which it is made are broken down – in huge towers – and reconstituted to make the artificial substances which are plastics. The petrochemical industry, which does this breaking and re-making of the chemistry of oil, and its mixture with any other chemicals one likes, has by now become indispensable to advanced societies. It provides everything from the nose-cones for moon rockets to domestic saucepans. (Actually, in this example, the nose-cone for the rocket was developed through the search for an oven-dish that was light enough to handle easily but tough enough to stand directly on the gas. It shows the flexibility as well as the vast resources of the petrochemical industry that it could switch in this way from consumer marketing to the moon programme, and realise that the same methods were applicable to both). The wide range of uses to which plastics are put is obvious in everyday life, but it is worth pointing out that they have by now become a substitute for many other kinds of natural materials which simply would not exist in large enough quantitites for all the demands that rapidly growing and very complicated societies now make. In this sense, oil is one natural resource that is being used (and perhaps used up) very fast, but it also implies a saving on many other kinds of resources.

One of the chief of these is the natural fibres of which all cloth was made before the Second World War. Oil is now widely worn as clothing, in the particular form of petrochemical product which is known as man-made fibres. Since the introduction of nylon, more and more of these have been produced, and they are very much cheaper to produce and prepare than the natural materials – linen, cotton, wool, silk – on which the world relied until so recently. By now, if supplies of oil were cut off, millions of the richer inhabitants of the world would be practically naked. The richest would naturally still be able to afford natural clothes, and the poor, in Africa and Asia, still often wear them. But, in between, the less rich inhabitants of the richer countries would again realise how vulnerable they really are. Here, too, it is true that coal and some other deposits of pre-historic life, *could* be used to make artificial materials, but, once more, it would be a very expensive process. Not only that, but not enough would be available if it were to be used to provide energy as well. Even more important than fibres, though, is the case of fertilizers. Nitrogen in the soil is the basis of most fertilizing agents, and on them depends the survival of millions – perhaps even the future of war and peace and the survival of the human race. In their turn, these depend on oil. It is practically impossible to overstate the importance of oil in this respect; and the rise in the price of the fertilizers which has taken place recently, in response to the rise on the price of oil itself, is little short of catastrophic. (I shall consider this more closely when looking at the dependence of the poor countries upon oil.)

In a minor way, oil can also be eaten itself. Since it consists of the corpses of millions upon millions of small organisms that died in a remote geological age, it is rich in food-value. It has been extensively used to feed cattle, (and therefore indirectly humans) and this use might spread a lot further. For again, it is cheaper, at least potentially, to transform oil into protein fit for human consumption through the medium of a cow than it is to transform grass by the same means – though one should add that while oil provides excellent meat, grass provides very much better milk. Nonetheless, grass takes land to grow on, and as demands on land will grow through the growth of human population, and demands for cattle grow for the same reason, so oil is increasingly likely to be used for food.

Finally, and perhaps most obviously, oil is a lubricant, and again it is often cheaper and more efficient to use it than older forms of grease which kept machines from seizing up before.

This very wide range of uses, in a cheap and efficient

form, is one of the three principal reasons for the general cheapness of oil. It is worth emphasising again that other substances could be used in every single way that oil can; but it is the fact that a tanker-full of oil can be put to so many different purposes, which makes it possible to base whole industries on the *same* processes and the *same* material. Obviously, this means that it is far better to use oil than to duplicate all the different forms of expenditure over the whole range of different kinds of plants and machinery that one would be obliged to if one were to use other substances instead. In saying this, I am overlooking for the moment the fact that there are of course different qualities of oil, and that not all of them can be used for all the different purposes I have outlined here; but even so, the same kind of process can be used all the time for getting it out of the ground, transporting it and delivering it, and that represents a vast saving.

The second major reason for the cheapness of oil is that it actually is so cheap to get out of the ground. Obviously, the costs vary. It depends where the oil deposits are, how difficult they are to get at, in what country they are found, with all the implications that suggests for alternative uses of land which would have to be sacrificed, and so on. Clearly, oil from the seabed demands an enormous expenditure on developing the new technology needed to extract it and the setting up of very expensive forms of plant indeed. But if one takes the most straightforward case, that of a country where the oil is easy to get at, where the deposits are large enough to make it worthwhile to set up the machinery in the first place, and where the land is mainly desert that does not have to be sacrificed for other purposes, then the cost of oil is astonishingly low. One such country is Saudi Arabia, and even after all the inflation of recent years, the cost there, even in 1974 of getting one barrel of oil out of the ground is ten cents. Cents, not dollars. It also so happens that Saudi Arabia has the biggest known deposits of oil in the world, and that most of these are easy to get at, and that they are of high quality. Which accounts for the great importance of Saudi Arabia in the politics and economics of the world's oil; but that is another question which will be looked at later. For the moment, the essential point is that oil (with its cousin, natural gas) is so far cheaper than any other deposit in the earth that no rival is even remotely in sight. And *that* is worth remembering whenever one reads about the possible use of new sources of energy.

For both these reasons, then, the fact that it can be put efficiently to such a wide range of uses, and the fact that it can be raised so much more cheaply from the ground than anything else, oil is by far the most economical source of energy there is. The third reason that I mentioned before, that it is comparatively easy to transport, is one that is probably best left for later consideration. This is because the costs of transport vary a great deal; partly for natural reasons like the difference between large and small tankers, or between the costs of pipeline transport and sea transport anyway; but also because the costs of transport have not always been reflected in the cost of oil to a country that imports oil. In Western Europe for instance, the costs of oil included, until fairly recently, a completely theoretical cost of transport which was fixed by the big international (but mainly American) oil companies, and which was actually a good deal more expensive than the real cost. The transport element in the costs of oil therefore depended, not on the real cost of transport, but on the working of the big companies, and this is something I will look at later.

But even so, it can perhaps be taken for granted that in questions of transport, oil has significant advantages; and this, taken together with the other considerations involved, has meant that one government after another has come to base more and more of its economy on oil. In fact, oil has now become the biggest industry in the world; and as for the world's dependence on it, a few figures will show how that has grown.

The most spectacular period of growth has been since the Second World War – the very same period that has seen the great economic miracle discussed at the beginning of this chapter. The two are very closely connected; and during that period, the growth in the oil industry has been bigger than the growth in any other big form of economic activity. World figures of production give a clearer picture than words. In 1945, the total output of crude oil (that is, oil taken straight from the wells, and not yet refined or adapted to any other uses) was 250 million tons. *Within five years,* it had doubled; and ten years later, by 1960, it had doubled again to 1,000 million tons. By now, the rate of growth was so fast that it became a question of how long it took to reach the output of an extra five hundred million tons again; that is, once more double the total production of 1945. Considering that the Second World War had been fought on oil, and that a great many forms of economic activity, especially in the United States had actually grown during the wartime period, the idea of a world using up twice the total output of 1945, and then doubling that and then adding twice the 1945 output to that again gives some idea of both the scale of the economic miracle, and of its dependence on oil. In fact, the next five hundred million tons were reached in 1965, and by 1968 an extra five hundred million was added to that. By the time the Russians were invading Czechoslovakia, in tanks and aircraft running on oil, shortly after the Middle East War of 1967 had been fought on oil, the world's output was 2,000 million tons. By this time, one might think, the rate of growth was bound to slow down, but it hasn't. Today (again 1974), the world output is nudging 3,000 million tons, and would in fact have reached and even surpassed that figure, were it not for the political difficulties surrounding the Middle East War of 1973. And most people agree that both the world's demands for oil and the world's production of oil will continue to grow so fast that by 1980, 4,000 million tons will be produced and used up every year. In other words, the most spectacular period of growth might have been the period since the Second World War; but if we go on as we are, the fastest growth is still in front of us.

At which point, one stops absorbing figures which in any case boggle the imagination, and begins to think instead of whether any country's economy can afford to go on growing so fast; and whether the new prices which are going to be charged for oil for the foreseeable future can be paid by anybody; and whether the effect on the atmosphere and the oceans of discharging ton after ton of sulphur from the burning of oil is not intolerable. Some of these questions will be considered later. Meanwhile, it is necessary to break the figures down somewhat, and consider to what degree different countries or areas of the world have come to depend on oil; and especially which countries can look after themselves in their use of oil and which can not. The first country to make extensive use of oil in its economy was the United States, well before the Twentieth Century; the second was the Soviet Union; the third area comprised Western Europe, and also Japan; and the poorer countries of the world are today trying to develop their own industries very largely on the basis of oil. Now that I have suggested why this general reliance on oil was practically bound to come about, and shown something of how much it has grown, I shall consider these different areas one by one, in the next chapter.

2

WHO USES OIL

THE UNITED STATES

The difference between the United States and other industrial countries is that the United States practically discovered the uses of oil. This is partly because there is a great deal of oil in the United States – indeed, even today, it is still the biggest oil-producing country in the world. (That is not the same thing as saying that it has the biggest known deposits of oil. Those, as I have remarked, are in Saudi Arabia.) But in terms of which country produces most, the record has been held almost continuously by the United States, though for a very short time it was overtaken by Russia. Clearly, most countries tend to make use of what natural resources they have, or did so until very recently, when the world's economy as a whole became sophisticated enough for some of them to switch from what they actually had to what was actually cheaper. And in this sense, the United States made wide use of oil for the same reason that Britain pioneered the use of coal: there was a lot of it there.

But this only suggests half of the peculiar position of the United States. It is not merely the world's biggest producer of oil, nor is it without challenge the oldest. It is the Russians who claim to have sunk the world's first oil well, and they too were pioneers in its use. But the Russians, around the turn of the century, did not have enough money available, nor sufficiently advanced technology to exploit it in the manner of the Americans. And this points to the second reason for American pioneering in the field of oil: by the end of the last century, the United States was already a very sophisticated industrial country. Gladstone, for one, had already remarked that it was bound to become the world's leading power in the twentieth century. In this respect, the United States had a sufficiently advanced and sophisticated society by the end of the nineteenth century to begin the process of exploiting and using oil for many of the multifarious purposes which the world has by now come to take for granted. More, since it was a society dedicated to the ideas of free commercial enterprise, the discovery of the uses of oil coincided with the rise of the gigantic American oil companies – notably Standard Oil, the empire of John D. Rockefeller. So the second respect in which the United States is unique follows from the first: not only did it get there before any other country, but it also produced most of the world's leading oil companies. Today, these dominate the international oil industry. But the third respect in which the United States is different from most other countries, is that it was for most of its recent history self-sufficient in oil. It produced enough for its own purposes, and indeed until the end of the Second World War, it was not only the world's biggest producer but also the world's biggest exporter. Before 1945, by far the greatest proportion of the oil products that Europe imported was imported from the United States.

Now this peculiar position has certain implications. It implies first that the great American oil companies were originally concerned with the exploitation of *American* oil, and its sale to the American public. Second, it implies that for much of the early period in the history of oil (which, it should be clear by now, really goes down to 1945) the European countries tended to import, not oil itself, but oil products, and that they did so largely from the United States. This, in turn, means that the American companies established international *markets* before they established

international *operations.* (In other words, they became companies that operated in other countries, not because they began by raising the oil in other parts of the world which possessed it, but because they sold oil products to other countries which had none.) Lastly, it implied that American companies had to worry about their own home market when it was discovered:
a) that the United States was no longer self-sufficient and might have to start importing oil – simply because its own economy was so advanced that it demanded more than anybody else; and b) that oil produced elsewhere (Mexico, Venezuela and then the Middle East) was actually cheaper, even if one took the transport costs into account, than much American oil. When this became apparent, the American companies could do one of two things. They could either move into fields abroad, and continue to make profits there; or else they could try to restrict the imports of oil into the United States, and so protect their profits at home. In the event, they did both. The fact that they did both lies behind many of the difficulties which exist today between countries which produce oil and countries which use it. But equally, the United States is still far less vulnerable to these difficulties than most of its friends and allies. For it is true that by now the USA imports oil at an enormous rate, but it does so because of its very high standard of living and because of a sophisticated – not to say profligate – economy. It does not depend on imported oil for its survival as many other countries do.

Altogether, the United States by now imports some 35% of its energy requirements – but this is to take all forms of energy into account. Considering that by itself, it consumes more than a third of the world's energy altogether, and that if it continues at the present rate of expansion, it might consume half the world's energy before the end of the century, this makes the country a vast importer of energy in every form. But less than half of these imports take the form of oil, and of that half only a proportion comes from the area which has led the revolt of the producing countries against the consumers – the Middle East. In fact, only about 6% of American requirements come from this politically sensitive area. Again, 6% represents a great deal by anyone else's standards; but by American standards it is clearly not vital. It has been calculated, for instance, that if American homes and offices, which are normally heated during the winter months to something like 70° Fahrenheit, and reduced to an almost chill level of air conditioning during the summer, were to turn their thermostats down by three degrees only, the country could save enough energy to make it independent of that 6% of imports. (Obviously, this would also demand other forms of adaptation, in terms of what kind of energy you use for what purpose, but all the same the global saving would be made.)

So the United States is not dependent on energy imports in the same way that other countries are; and it is particularly not dependent on imports from the Middle East, to which other countries are most vulnerable. Yet as a result of its early expansion in the field of oil, it dominates most of the international oil business. The consequence is that American *companies* have been very sensitive indeed to what goes on in the politics and economics of oil; but that the American *economy* allows the American government considerable room for manoeuvre in a crisis of the kind that happened towards the end of 1973. On the other hand, the European and Japanese *economies* are bound to remain highly sensitive to any event which threatens to cut off their oil supply, but (with the exception of British Petroleum and the Anglo-Dutch company, Shell) have a relatively minor part to play in the international oil business.

Equally, the countries which actually produce the oil have to deal with American companies (and have often had to deal with the American government) if they wish to change the terms on which oil is supplied, or to restrict or increase the supply of oil; but the results of their dealings do not affect the American economy overmuch; they affect the economies of other countries. In between, the companies continue to enjoy the prospect of the gigantic profits which are made from extracting the oil in one country and selling it in another. This puts them in a very strong bargaining position, since they can switch around a lot between suppliers of oil and customers for oil, and such a world-wide network of operations is beyond the power of most governments to control. One result of this is that recently some of the states which import oil have tried to deal direct with the states that have oil to export – but more about the operation of the companies and the new relations that have arisen in later chapters.

For the moment the point is that it was the American position as the original exporter of oil products which encouraged the American companies to expand in the way they have; and that the continuing American position as a country which needs to import less oil than most industrial states gives it a peculiar strength in dealing with other governments.

That does not mean that the United States has no problems. The fact is that very shortly after the Second World War, it became apparent that there were going to be fairly severe restrictions on the further exploitation of American oil. This came about either because there were physical limits on how much one could reasonably get at and still make a profit, or because certain commercial bodies in the country kept a tight watch, for commercial reasons, on the amount that was being produced and sold. This fact coincided with the discovery that the reserves of oil elsewhere in the world were so vast that they could go on being used for the indefinite future. (Or so it seemed at the time.) Together, these restrictions and these discoveries meant that it was going to become very much cheaper to import oil to the United States than to use more American oil. It all happened so fast that by 1948, Americans were importing more oil than they exported (in other words, – they became net importers). And the growth in the imports of oil over the next ten years continued at such a rate that in 1959 a limit was fixed by President Eisenhower on the amount of oil that could be brought into the country (that is, an import quota was established). This was in the beginning no more than an attempt to protect the American companies in their own home market, but it was to have dramatic consequences for the future, which will become clearer when we have considered the relations between the oil companies and the countries which produce oil. In the meantime, the fixing of quotas has ensured that the United States continues to produce much more of its own oil than it would otherwise have done, for in America oil can be very expensive indeed to lift out of the ground. Compared with the cost of ten cents a barrel in Saudi Arabia, the average cost in the United States is now well over three dollars. This has meant that, until very recently, Americans were paying more for their own oil than they would have done if they had imported other people's, even when all the additional costs for oil from abroad are taken into account.

As a result, American self-reliance in oil has proved expensive. *But* it has meant two things. First, that the United States, though one of the world's biggest importers, is politically more powerful than the others. Second, that it gradually became actually worthwhile for the American companies to look for new sources of oil inside their own country. So here the wheel came full circle. Having begun to import because it was, among other things, too expensive to develop more American oil fields, the country then set restrictions on imports in order to protect the home companies; and having done that, it found itself paying a high enough price for its own oil to make it worthwhile to develop more oil fields after all. The result is that the United States has now begun to exploit new fields even in areas as forbidding and as prohibitively expensive as Alaska. It has also begun to develop other sources of oil which in the past were not considered worth looking at. These facts – the new fields and the new techniques – are among the considerations which today persuade many governments, not only in America, that the present great advantage held by the oil-producing Arab states in the world markets will not last more than ten years or so.

At the present moment, therefore, as at the beginning, the United States maintains its position as the world's biggest producer of oil, though it should perhaps now be clear that this is a somewhat more artificial position than it might appear to be . Equally, it is still a very big importer of oil. But meanwhile its importance as an *exporter,* even of oil products, except in some areas of advanced technology, has declined. Today the chief friends and allies of the United States import their own oil requirements, not from that country, but from others, mostly in the Middle East. These other partners in the gigantic economic miracle of the last generation have based more and more of their economies on oil, and imported more and more. They are, primarily, Western Europe and Japan.

WESTERN EUROPE

The countries of Western Europe, to take them first, were comparative latecomers to the use of oil. There were obvious reasons for this. The Industrial Revolution occurred first in Britain on the basis of energy derived from coal, and the pattern of industrial development was exported from Britain to other European countries which also had large coal deposits – Germany, France, Belgium in particular. Not only that, but they didn't have any oil.

Since they appeared to be able to manage without, the real question was what prompted them to search abroad for oil. And at least one major answer is that it was organizations operating abroad: the European navies. Clearly, this was not the only reason, since well before the First World War Standard Oil was exporting kerosene to Europe for oil lighting, and the car was making its appearance on European roads and thereby prompting a demand for petrol. As a result, Shell, which got into the oil business in the first place by transporting Russian oil from Baku, was fighting an economic war with Standard Oil by the turn of the century, and shortly afterwards merged with Royal Dutch to improve its competitiveness. Also in consequence it searched for new fields. But the real stimulus to the search for oil outside Europe was naval demand – since ships, like most other things, can run more cheaply and efficiently on oil than coal. The result was that European companies began to exploit and open up oil fields – at Brunei in what was then the East Indies, and nearby in what was then the Dutch East Indies (today Indonesia); in parts of what was then the Turkish Empire; and in what was then known as Persia; and they did so to an extent larger than the actual European demands for energy from oil would by themselves have warranted. This was important, because for many years thereafter a great deal of oil was being produced abroad but at the same time, even with the growth in the use of cars for transport, even with the fighting of the Second World War on oil, even with the widespread use of petroleum products, European energy needs were still largely met from coal. It was this situation which led to the sudden emergence of oil, as not only *the* cheap source of energy but also as a kind of world-wide surplus later on, and prompted European governments to base more and more of their economies on oil – to adapt to the use of oil rather faster than they might otherwise have done, run down their own assets – like coal-mines –

in the process, and so expose themselves to being very vulnerable indeed when a crisis in the supply of oil threatened.

But before that time was reached, another factor had also given a large boost to the Western European demand for oil. This was simply the destruction brought about by the Second World War. So many coal-mining areas were devastated by the war that by the end of it Britain – even in its run-down condition – was producing 60% of all the coal in Western Europe. This was a real energy crisis, and to meet it, the countries of Western Europe began to import large quantities of oil from the United States, Venezuela and the Middle East.

Now at this point, it is important to bear in mind another fact. Before the Second World War, not only had coal been the prime source of energy for Western Europe, but because it had fulfilled that function, the real demand in the European continent was not for fuel oil (which is burned directly) but for oil products. This, combined with the fact that at that time ships, including oil tankers, were not particularly big, had meant that the cost of importing oil in its natural state – crude oil – was too high to be economic, and it was cheaper to import the products. Oil, as a result, was refined into its other products first, and then imported into Europe. But now a change happened. First, it became more expensive than the European economies could bear to import the products rather than the oil, since the products had to be paid for largely in dollars, and Western Europe at that time was not a good dollar earner. Indeed, most of the dollars the countries had available came from American aid anyway. Second, the extension of pipelines and the building of larger tankers meant that it was getting cheaper all the time to import oil direct and refine it in Europe. As a result, European governments began to encourage, or pressure, the oil companies to build oil refineries around the coasts of the continent, and that meant that increasingly large supplies of oil were beginning to arrive, comparatively suddenly, to be adapted to any purpose the governments or the importing firms wished. Above all, it meant that large supplies of fuel oil were now reaching Europe for the first time. As a consequence of fuel oils becoming available, the demand for oil in Western Europe went up by leaps and bounds.

There were good economic reasons for this. In a period when the Western European countries were still very short of foreign currency, it made sense to save as much as possible by importing the oil rather than the product, and indeed by the mid-1950s, these countries were saving something like £50 million a year in foreign exchange, purely by using their own refineries. But it also meant that once you had the oil, and provided it remained cheap, you might as well use it in preference to other sources of energy which cost more. And oil *did* remain cheap – largely as a result of the surplus which earlier European policies had helped to create, and it looked like remaining cheap for a long time, largely as a result of the artificial surplus which the American limitations on imports were also helping to create. It therefore seemed sensible to rely more and more on oil for energy; and it *was* sensible so long as one looked at the question purely in economic terms. Once you had built the refineries, it was money well spent; why, then, go to the additional cost of modernising coal-mines, or opening up new ones, to provide a fuel which would still be more expensive anyway? And in the mid-1950s, other arguments hardly mattered. It did seem that in political terms, most of the countries which produced oil in the Middle East were likely to remain friendly to, or dependent on, the West, so there seemed to be little political threat to a secure supply of oil. Nor did there seem to be any real military threat. The Soviet Union was hardly in a position to take over the Middle East; and in any case much of that area was bound by a network of alliances and treaties with the United States and Britain, so the Russians would know that they risked a world war if they tried to move in. If, therefore, neither the political nor the military arguments seemed to matter, the economic argument was bound to predominate; and the economic argument was decisive: oil was cheaper.

So, throughout the late 1950s and into the 1960s, oil began to replace coal as a source of energy in Western Europe at a faster and faster rate. By 1966, oil was the most important source of energy in Western Europe taken as a whole; and by 1970, this was also the case in every single West European country taken individually. It is worth remembering that all this happened not when the decision to close down the mines had been already taken, but in direct competititon with attempts to keep the mines open. It is just that the economic arguments for the use of oil always won; and one result has been that many countries, but especially Britain, have imposed taxes on the use of fuel oil which have been raised solely for the purpose of subsidising coal-mines, and preventing the distress to the mining areas which would have resulted if the switch to oil had happened *too* cheaply and quickly. These taxes have in fact been quite low – only a few pence on the gallon; but in terms of the actual cost of oil, they have been pretty high – nearly fifty per cent of what a gallon cost when it left the refinery. The result has been that coal mines survived better than they might otherwise have done, and this has meant that there is a certain ability in Western Europe to turn back to coal, at least in part, if supplies of oil seem to be in danger. It has, of course, also meant that coal miners now represent a very important actual and potential force in the economy, and has strengthened their hand greatly in bargaining with governments. A change from the long years when they were made painfully aware that they represented a weakening force, to whose views one did not have to listen overmuch.

But whatever the shape of future energy requirements and their sources in Western Europe the present situation is that of an oil-based economy. Here too, the figures speak for themselves. In 1962, Western Europe consumed nearly 264 million tons of oil. Ten years later, it consumed 704 million tons. In 1962, oil was still just under half of what Western Europe needed for its total energy consumption. By 1972, it was 60%. Since then, the use of oil has continued to grow.

So far, though, I have spoken about Western Europe as if it were all of one piece. But in fact there have been some rather important differences in the way that the various West European countries have handled their energy policies. On the whole, West Germany, Italy and the Scandinavian countries have adopted a policy of trying to bring the price of oil down as far as possible, (which has meant a conflict with the pricing system of the American oil companies, which will be examined in more detail later on) whereas Britain, as the home or headquarters of two of the world's biggest oil companies has tended to see an advantage in maintaining rather higher prices. Equally, Britain was dominated by the big international companies more than any other European country, and this has meant roughly that Britain has paid more for its imports of oil than any other European country. (Indeed, in the mid-1960s it was a quarter as much again.) The connection between this fact and Britain's continuing record of difficulties with its balance of payments and its economic development is rather obvious – though it does not appear to have been so to the people who made the decisions in the City and the government. (One day, they might sort out Britain's real interests from its bogus interests.) And France has followed a different pattern again. For in France, a main consideration was to try to import oil from the franc zone, which meant principally Algeria. This was no doubt partly to help to keep the franc zone alive, but it also meant fixing prices at a fairly high level in order to make it worthwhile to develop Algerian oil. This didn't do the balance of payments any harm, so long as Algeria was part of France anyway, or stayed in the franc zone while that was still alive; but it did mean higher oil and petrol prices in France. And that in turn meant that the French state took a very active interest in the operation of the oil business – in order to keep the international companies from under-cutting on the home market. The French state company, the CFP (Compagnie Française des Pétroles) has therefore tried to serve the French government's interests more closely and directly than any other major oil organisation. Even more, in fact, than the Italian state company, ENI, which was more like an attempt by the Italian government to open up a line of direct competition with the major companies in the Middle East.

These variations in national policy in Western Europe partly underlie the difficulties which the European Economic Community now has in trying to establish a common energy policy for the Nine. But an even more important cause of these difficulties is the fact that in the years leading up to the establishment of the Economic Community, oil was not a very important feature of European energy policy. After all, the first major step towards the Community was taken when the continental governments of Western Europe established the European Coal and Steel Community. This has meant that they (and especially France and Germany) could agree on the important decisions about the production and uses of the prime source of energy and the prime material of industry – which they were in 1949 when it was all set up. At the time, it would have been pointless to talk about a common energy policy: one was already implied in that first step. But as coal has been replaced by oil as the major source of European energy, so it has become rather embarrassingly clear that the different European countries have different policies towards the uses and prices of oil, different relations with the states that produce the oil, different attitudes to American policy, and different interests in the terms on which they can go on buying oil. Out of all these differences, they are supposed to make a common energy policy, and the West German government in particular has taken the line that the European Community can not survive without one. But so far, nobody has shown that they even know how to begin; and for much of 1974 there was a serious danger of the kind of competition that I mentioned at the beginning; an attempt by individual countries to outsmart the others in dealing with the oil-producing states. This is not only the opposite of a common energy policy, but a serious cause of political disagreement within the countries of Western Europe.

So where, in the end, has Western Europe got to? It now has two main sources of energy: coal – of declining importance – and oil. But it is also increasingly developing a third; natural gas. In these respects, it will be rather highly favoured among the nations of the world by the end of the present decade. It will certainly still largely rely on oil for its requirements, but it will have the chance either to supplement this, or else to begin to replace it, either through going back in some measure to coal, or else by developing gas resources. But while this is a great advantage in theory, it also means that any rational decisions will have to be taken on a European basis rather than that of any individual country. The way that the different European states are going about it at the moment does not suggest that they will succeed in creating either the will or the institutions to form a common energy policy in the foreseeable future. And yet, they have so many options open. They can continue to depend for a fair proportion of their needs on coal or even go back to coal for a higher proportion if they want to. They can switch from coal and/or oil to gas. And they can do something else which has not been mentioned yet: they can produce more of their own oil themselves.

This last choice is one that has appeared increasingly open to them as several Western European nations have found more and more oil fields in the North Sea. Britain, the Netherlands, West Germany, and Norway (which is not a member of the European Economic Community, but which is likely to become an important exporter of oil to the Community countries all the same) have all found important deposits of oil in the North Sea. And some, notably Britain, have found very considerable sources of gas as well. On the other hand, France still attaches great importance to the importing of gas from Algeria; and yet Algeria is an oil-exporting country too. As such, it plays a considerable part in the counsels of the Arab oil-states, and in this sense France (and the countries which rely to a degree on French imports of natural gas) are not so free simply to substitute gas for oil as might appear at first sight. So, in the end, the Western European countries might look as if they are in a relatively good position to develop their resources and expand their economies without too great a dependence on oil. But the very fact that they have such a good range of choices, and that these choices demand a lot of co-operation between them, means that they will either have to frame a common policy or else, probably, go on paying what they have to for their oil imports. And therefore one of the chief questions has become: how much of Western Europe's oil needs can be met from the North Sea?

Here, the answer is uncertain. It depends on three things. First, the size of the North Sea fields. Second, the choice, which is still open to individual countries as to whether they should in fact export their "own" North Sea oil to their European partners. Third, what kinds of uses the oil is suitable for. To take them in reverse order: it is fairly clear by now that most North Sea oil is likely to be quite reasonable as a source of energy, but not altogether suitable for conversion into petrol. For that reason, Western Europe is likely to go on needing to import large quantities of oil, however much it finds under the Sea. The answer to the second consideration will very largely depend on whether it is possible for the European Nine – the members of the Community – to create an agreement on energy policy. If they do, they might well make use of their off-shore oil themselves, and so reduce their energy bills dramatically. If not, separate countries might sell it in the best market they can find, anywhere in the world. And as for the first question – how much oil there is actually available under the Sea – nobody knows. Estimates have constantly been increased ever since the first drilling began; and known deposits are growing all the time. But this does not mean of course that they will keep on growing. The limit will be found some day and it might already have been found. But if one takes the three questions together, a reasonable guess might be that North Sea oil will start making a significant difference to Western Europe's energy imports by about 1976, and that after that, for at least twenty years, it could supply about 40% of *all* the region's oil requirements. Taken together with the prospect that gas from the same source could also fulfil a large part of the overall energy needs, this might seem to promise a Western Europe very largely freed from the dependence on the imports of Middle Eastern oil which grew up between the mid-1950s and the late 1960s.

But two points have to be born in mind. The first is that such a degree of freedom will depend very largely on the ability of the different European governments to work out a sensible energy policy between them, which would imply a sharing-out of the tasks appropriate to different kinds of energy across the Community as a whole, and also a policy on European money which would mean that no one country stood to make profits that were too large at the expense of its neighbours. And, second, it would still mean that the energy obtained in this way would be more expensive than oil imported from the Middle East. The costs of getting oil out of the North Sea are very high. Even now it is doubtful whether it would be available for a long enough period to cover these costs completely, when they are compared with the costs of getting oil from elsewhere. So that even if a common policy were agreed, and even if Western Europe became less vulnerable in terms of its *political* relations with the outside world, it would still have to pay more than it does now, or has been in the habit of doing, for its basic sources of energy. And this second fact in turn makes it less likely that the governments of the different West European countries can reach a common energy policy; for in the meantime, some of them will, again, be tempted to reach relatively advantageous agreements with other countries that produce and export the oil: agreements which might not cost them too much in terms of hard cash but which can be carried out in technical aid; and this temptation will be particularly strong in the middle years of the 1970s, when the North Sea oil has not yet begun flowing ashore in large quantities, but when the European countries still need large quantities of oil because they have made themselves so dependent on it through their earlier policies.

This might have important implications for the relations between the European and the Arab states. One way of trying to regulate these relations is the establishment of what has come to be known as the "Euro-Arab dialogue", which is really no more than an attempt to improve relations between the Economic Community and the Arab countries, especially those on the Mediterranean. In return for what Europe still needs so badly – oil – it can also mean an attempt to explore with a little more care than in the past what Europe has to offer besides money. In this way, the "dialogue" might become the framework for a whole programme of economic and technical collaboration, and could greatly speed up the pace of that social development which is the declared aim of most Arab leaders. All the same, one has to admit that so far it is very little more than talk – mostly French – and that very little has been done to work out a real programme.

That might change. Meanwhile, the European leaders are acutely aware that although in the long run their demands for energy will be met from a variety of sources and their general position will be very flexible, they are in the short run extremely dependent on oil, and Arab oil at that. And here, too, the very fact that they are now paying much higher prices for oil has meant that they have less money available to invest in a more flexible economic development over the next few years. So the position will not change rapidly, even when North Sea oil starts coming ashore. But Western Europe, dependent though it is, is in nothing like the position of the other great partner in the world's economic miracle of the past few years – Japan.

JAPAN

Japan is now the third economic power in the world, and easily holds the record for the longest continuous period of fast economic growth since the Second World War: some ten per cent a year over several years. Indeed, if it kept up this rate, it could soon become a bigger economic power than the Soviet Union, and it is just possible that it could overtake the United States by the end of the century. Until a short while ago, foreign observers were fond of predicting that the country *would* keep up that rate of growth, but it has now slowed down considerably, and a burst to the front of the field by the Year 2000 is now very unlikely. The reasons for the slowdown are very closely connected with uncertainty over Japan's supplies of energy, and in particular of oil.

The first thing to appreciate about the country's astonishing economic development is that it was achieved on fewer natural resources than those of any other industrial power. Until the Second World War, Japan did have just about enough coal to modernise and expand, but even then, this was only because a large part of the population was still living the traditional life of Japanese peasants, which meant that they endured conditions of great cold every winter – far more than Europeans could find tolerable – and did their own domestic heating on wood. Japan also harnessed the natural energy from its many mountain rivers in the form of hydro-electricity. During the war, it of course needed much more oil than before in order to keep the war machine going, but this was largely supplied from the oil fields it captured in Burma and the East Indies. But at the end of the war, it was left with an old-fashioned, and, in places heavily damaged, coal industry; with a need for rapid economic development; and with a series of social changes which meant that a large part of the population was going to move into the cities and need rather more sophisticated heating. In other words, Japan faced the problems of Western Europe after the war, but in a much worse form, and since it was a vanquished nation (unlike Western Europe which contained both victors and vanquished) it had more limited room for manoeuvre and even more severe problems in getting hold of usable foreign currency. The result was inevitable: Japan, too, needed oil and needed crude oil that it could refine itself, rather than buying oil products from abroad. But because of its very weak position, it accepted terms for the building of domestic refineries which were very different from those in Europe.

In effect, what happened was that Japanese companies put up half the cost of the refineries, but the international oil companies put up the other half, especially in the foreign currencies which Japan did not have. In return, the international companies secured the exclusive right to supply these refineries with crude oil for ever. The result has been that in later years, Japan has been forced to pay higher prices for its oil than other countries importing from these same companies, for in Japan they faced very little competition – they supplied about 80% of the country's oil imports throughout the 1960s. Not surprisingly, the Japanese government has tried to cope with this situation by two measures. It was tried to find supplies of oil from elsewhere, to reduce the importance of these refineries and their imports; and in order to ensure that *other* refineries could be built on different terms from those it had accepted in the first place, it has taken over more or less direct control of all Japanese ventures in the field of oil.

But before considering what this has come to mean, it is worth looking first at the sheer volume of Japanese oil imports. Here, too, it has faced problems like those of Western Europe, but in worse conditions. For the Japanese coal industry is very unfavourably situated, the coal is extremely hard to get at, with the result that it takes much more manpower to lift a ton of coal out of the ground than it does in Europe, and in consequence it is extremely expensive. (The fact that comparatively cheap coal was not available of course also encouraged the oil companies to charge higher prices in Japan, since it meant little domestic competition from other fuels either.) All this taken together meant that Japanese oil imports rose at a staggering rate. In 1950, when the first of the new refineries were built, oil supplied less than one-twelfth of Japan's energy needs. Ten years later, oil was of equal importance to coal in the economy – and, alarmed at the clear prospect of dependence on outside supplies for the greater part of its energy, the government had already tried to discourage people from switching to oil; but it was too late. By that time the expense of coal was growing fast, that of oil declining somewhat, as other companies moved into Japan to offer rather cheaper oil for new refineries. The result was that from 1960 to 1964, the use of oil in the country rose by 25% *a year.* By 1972, oil had risen from its 1950 position to reach 76% of the energy consumed. Today, just as Japan is the third largest economy in the world, so it is also the third consumer of oil; and unlike other countries it has very little alternative available in the form of either coal or gas.

Japan is thus more dependent on oil than any other country in the world, and more vulnerable to a break in supplies. Which is where its government's policies of trying to ensure several different sources of supply come in again. It has tried to do three things. The first is to look for oil in the seas around Japan. So far, this has not been very successful, although there are good prospects of strikes farther afield, notably in the South China Sea and in areas near the Philippines. The trouble here is that it demands agreement with other governments who might also stake their claims in the area, and also – possibly – with the oil companies who might try to exploit it too.

The second is to look for an alternative supplier, in this case the Soviet Union. Here, some agreements have been reached, which should mean that the Soviet Union will shortly be supplying about 15% of Japan's oil requirements; a hefty proportion, but not one which would dramatically reduce Japan's dependence on the Middle East or the oil companies. The Japanese government also hoped, in co-operation with the Soviet Union, to put more emphasis on gas coming from Siberia, and so reduce its overall dependence on oil anyway. But to do this, it needed to offer capital to the Soviet government to open up the gas fields in the first place, and this it was about to do, along with some American companies, when Mr. Brezhnev suddenly decided that his country was not interested. Whether this will always be true depends on the politics involved; but for the time being, Japanese co-operation with the USSR looks as if it is likely to remain important but limited.

The third course of action open to the Japanese government is to try to by-pass its dependence on the companies and open up a more direct interest in the Middle East. This course of action in turn breaks down into more than one procedure.

One obvious way is for Japanese companies, or groups combining state and company interests, to try to open up new fields directly in the Middle East. Here, the start was not hopeful: one attempt was made at the beginning of the 1960's to open up the oil field at Khafji in the Persian Gulf. But the producers of crude oil from that field soon found that they could not really sell it in Japan, and were forced into the position where they had to offer a share in the company to the Americans in order to recover some of their costs; in return the Americans had the right to a fifth of the oil – so what was begun as an attempt to get more and cheaper oil for Japan ended up by providing more for American companies. Since then, though, things have improved; and the Japanese now play a very important part in the development of the extremely rich fields at Abu Dhabi, also on the Gulf. So here there are some prospects of success after the first setbacks – especially since the success was won in very fierce competition with the international companies.

A second way has been to try to force the international companies to back down from the highly unfair and profitable advantages which arose from the manner in which Japan was forced to accept the terms for their refineries in 1950. The Japanese government has sometimes tried this, and has to a degree been successful, but only when supplies of oil were secure. When supplies look vulnerable again, the government tends to be a little nervous about taking on the companies – and indeed the companies have on occasion made specific threats that if action were taken against them they would simply reduce the supply of oil. So that although the companies are not by now in the position to assure themselves of the hundred per cent supplies to the refineries which they originally had, they are still able to ward off further pressure, even when they face criminal prosecution under the country's anti-trust laws; and it must be said that in this, Japanese business has usually shown itself to be more reliable as an ally of the companies than of the government.

A third way in which the Japanese government can by-pass the companies is to try to do deals direct with the Middle Eastern governments. On the face of it, this is an obvious procedure – after all, the Middle Eastern governments have also shown some interest in breaking the power of the companies, or at least in forcing them to pay more. So a little healthy competition from a major customer like Japan should be in the interests of both sides. The difficulty, though, is that such an arrangement makes Japan much more vulnerable to the political pressure of the Middle Eastern states. And when one such deal was concluded early in 1973, it led very soon afterwards to a situation where the Arab governments were able to put very powerful pressure indeed on Japan to break off relations with Israel. Now, Japan might not care very much about Israel, but it does have its major ally, the United States, to think of at such moments; and the embarrassment of the Japanese, caught between these two sets of pressure, was acute. On the whole, since then, they have shown a greater disposition to work with the companies after all.

In general, Japan will remain very dependent on oil imports for its economic survival, let alone its prosperity; and for the foreseeable future most of these imports will come from the Middle East. Which means that this uniquely vulnerable country will go on trying to maintain good relations with both the companies and the governments operating in that region, even though it is the most spectacular case of all in the economic miracle of the Western world.

So far, I have talked about the three main regions of the Western economic system which has been built up since the Second World War. It should be clear enough by now that the great wealth created by this system has been more and more dependent on plentiful supplies of cheap energy; and that oil has provided for all the countries concerned, *the* great solution to the problem. Indeed, all the optimistic assumptions I mentioned at the beginning of this book about the prospects of economic growth for the future have depended on a kind of belief, very largely unspoken, that cheap energy would be available for years to come. Without it, not only do the prospects for continuing growth begin to look very much more doubtful, but the very economic system itself is in grave danger. It is in danger not only because more expensive energy means that everything else also becomes more expensive; but also because the payment of higher prices for oil means that vast amounts of money are transferred from the main industrial states to the main oil-exporting states, and unless there is some agreement on the uses to which these sums will be put, the money will become meaningless, trade will begin to break down, and there will be a gigantic slump. And ever since 1970 oil *has* become more and more expensive. The nightmare of the Western economists still doesn't have to come true (the "Euro-Arab dialogue" is one way in which it is hoped to prevent it from doing so) but it now threatens to.

Before going into this new state of affairs, though, and in order to understand some of its further implications, it is necessary to look at the dependence on oil of the two main other areas of the world's economy: the industrialised, developed Socialist countries led by the Soviet Union; and the great mass of poor countries which are still hoping to develop their wealth.

THE SOVIET UNION

There are two obvious points to make about the Soviet Union right at the start. They are that it has an old and very large oil industry; and that it is the oldest Socialist economy there is. The two points are more connected than they might appear at first sight, though, because the fact that the Soviet Union is the oldest Socialist economy has meant that it was determined to be economically self-sufficient in everything during the years of Stalin's reign; and this pattern is still largely true of its energy policy. It does not *want* to have to rely, in however small a degree, on imports from anyone else. And the fact that it has large oil fields means that it can hope to remain self-sufficient for some time to come.

The development of an oil industry in Russia began in the 1880's, when the countries of Western Europe were casting about for an alternative to the oil empire which had been built up in America by John D. Rockefeller. The first major Russian field at Baku near the Black Sea was easy to get to, it provided oil of high quality, and it was much nearer the rest of Europe than were the wells in the United States. All of which were good reasons for a Rothschild to invest money in the Russian fields or a Shell to construct the tankers to transport their products. Equally, they were good reasons for the Soviet government after the Revolution to make sure of controlling this asset, and trying to continue the pattern of export. This is important, because it meant that one of the chief interests that the Soviet government showed in oil once the damaged fields had been got going again after the Civil War, was to use them for export. The country itself carried on much of its own industrial development on the basis of coal and hydro-electric power; and a deliberate decision was taken to discourage the use of motor transport in order to cut down on the use of oil. (Considering the state of most Russian roads at the time and for long after, it made more sense to rely on railways and rivers anyway.) But the real point is that the Soviet Union emerged as an industrial giant without tapping its own oil reserves to anything like the extent that might otherwise have been expected. It was the Second World War which changed the pattern.

First, the war changed things because it prompted an immediate demand for more oil, and second because the oil-fields themselves fell into German hands. As a result, extensive searches were undertaken to find new oil-fields in Soviet territory, well away from the German armies. These searches were highly successful – though not in time to make much difference to the Soviet war effort. *That* was sustained very largely by Allied convoys. But shortly after the War, the Soviet Union began to exploit the newly-found fields in the area between the Volga and the Urals, and then in Western Siberia. The output from the Baku area and the newer fields has now grown so much that today the Soviet Union is the largest oil producer in the world after the USA and might indeed overtake it at times. This has meant that the Soviet government, as aware of the advantages of oil as anyone else, has naturally permitted the oil and natural gas found in these fields to assume a larger and larger share in meeting the energy requirements of the country: in the 1960s they began, for the first time, to meet more than half. By now the proportion is about 60%, and it will certainly go on growing as the pace of exploration and exploitation of new fields continues. (The pace is such that the Soviet Union could very well become and remain the world's biggest oil producer after 1980.)

But, all this happened within an economy which was still very largely based on coal and hydroelectric power; and more than this, it took a considerable time and a great deal of money to get the new fields working. In other words, it was not necessarily cheaper at first for the Russians to switch to oil, if one accepts their reluctance to rely on imports. And even when they did switch, they tended to prefer natural gas to oil. So, unlike the West Europeans, who were perfectly willing to import oil if it was cheaper, and who had access to well-established industries in the Middle East which had already been paid for in large part, the Russians were paying *more* to develop their oil industry in the first instance than if they had gone on relying on other sources of energy. The obvious way to recoup some of this expenditure was to export. And, because it was less vital for them, and because they wanted Western money anyway in order to buy Western goods of other kinds, they began, in the mid-1950's an energetic export drive in oil.

There was one other aspect of this export drive which is worth noting. After the war, the Soviet Union acquired a considerable empire in Eastern Europe; and, of the countries which came under Soviet dominance, only Romania had an oil industry of its own. The rest were expected – indeed obliged – to industrialise, but as they did so, their own needs for oil obviously increased. And although they were made to industrialise on the Soviet pattern at first, relying for their heavy industries on coal – of which there was a large amount in East Germany, Poland and Czechoslovakia – they too began to use diverse sources of energy later on, just as the Russians did. This has meant that since Romanian supplies are far from adequate, the Soviet Union has been exporting large quantities of its own oil to the industries of Eastern Europe, to keep its own system going. This is still the pattern today: the Eastern European Socialist countries import most of their oil from their Russian neighbour – often on terms which are very much more expensive than if they bought it elsewhere – but it is not yet certain whether the Soviet Union will have enough for its own needs and for those of the Eastern European states, as well as the amount it wishes to export, for very much longer. Some suggestions are that after about 1980, the East European countries will be running short of oil unless they import it on a much bigger scale from elsewhere. This might be an important element in the relations between Eastern and Western Europe in the future – assuming, of course, that the Western members of the Economic Community succeed in agreeing on a common energy policy by then.

Meanwhile, the Soviet Union has gone on exporting oil, both to Western Europe and Japan. It succeeded in doing so in spite of some vigorous attempts by the United States to persuade the Western Europeans not to buy during the height of the Cold War period – though one can't help suspecting that these attempts were not purely an expression of political concern, but had as much to do with the fact that Russian oil was a good deal cheaper (for the Western, not the Eastern, Europeans) than that supplied by the American-based oil companies. As so often in dealing with the Russians, Sweden led the way, but it was rapidly followed by Italy. Indeed Italy now gets well over 16% of its oil supplies from the Soviet Union, and has been doing so pretty cheaply. (It also has the added advantage that the Russians agreed to take a lot of Italian industrial products in return.) Thereafter, France, Belgium and West Germany all followed suit; and only Britain and the Netherlands have refused. To nobody's surprise these are the only two European countries which provide the headquarters of two of the large international oil companies.

The basic pattern that has emerged in the Soviet Union, then, is one which suggests that it will go on being self-sufficient in energy for a long time to come. Here it contrasts with the other giant, the United States, which no longer hopes to be self-sufficient, and is continuing to import large quantities. Second, the Soviet Union developed its newer oil industries not only for its own consumption but also for export, and has to a degree built up a reliance on Soviet oil in other countries. One of the main questions that the pattern raises is whether, if the rate of development in the Soviet Union continues at its present rate, and if the Russian standard of living goes on rising, the country will be able to go on exporting in the present quantities. If it cannot, the East European states are likely to feel the pinch first; but so might other countries. For the present, though, it is no use making predictions, since there are two very large unknowns in the situation. The first unknown is the size of the North Sea oil fields, which might make Western Europe very largely indifferent to Soviet oil – and perhaps even leave some of it available for Eastern Europe after all. The second is the size and availability of the great deposits which the Soviet Union is known to have in Eastern Siberia. (So far, it has exploited only the oil and gas fields in Western Siberia.) The difficulty with these fields is that, although they are known to be large, they will be fantastically expensive to exploit. Eastern Siberia is a very long way from anywhere at all, and the cost of getting the oil out, setting up the pipelines and other forms of transport, and refining it are probably more than the Soviet government alone could stand. But so far, it has not shown great interest in American, Japanese or Western European help. In other words, the potential worth of the East Siberian field depends on political as well as geological unknowns.

If this field is *not* adequately exploited, some observers have suggested that before another ten years are past, the Soviet Union might actually have to start importing energy rather than exporting it, as its own economy keeps on growing. To which possibility, some slightly more alarmed observers have predicted that in that case, the Russians might take an unhealthily close interest in the Middle East. Well, it's a possibility. But at present there is little to suggest that the Soviet government would find it of great interest to risk a major conflict with the United States in that area, nor that Iran or the Arab states would like to see themselves tied too closely to the Russians. What is perhaps more possible, *if* the Soviet economy keeps on growing at its present rate, and *if* the East Siberian fields are not exploited, and *if* Soviet exports to Eastern Europe also continue, is that the Soviet government might try to come to some partial agreements with the Arab states, and so make it more difficult for the Western Europeans or Japanese to meet all their oil requirements from the Middle East. But here again, so much depends on such a wide range of political developments in the meantime that it is pointless to make any prediction. It is worthwhile only to bear the possibilities in mind.

The Soviet Union and Eastern Europe, though their own industrial development has been outside the system created for the Western world, represent very highly developed industrial countries. The other parts of the world do not. They are still struggling to develop fast enough to catch up with their own birth-rates and create enough wealth to survive on. It has looked like a losing battle in the past few years, though not yet a hopeless one. But one of the severest blows they have had to face is the recent steep rise in the price of oil. It threatens to wreck their hopes and plans.

THE DEVELOPING WORLD

The point at which one must start in considering their problems is that these countries use very little energy. This might look like the end of the problem, rather than the beginning, since, one might ask, how can they be seriously affected if they are using very little energy anyway? But there are two related points to bear in mind. The first is that one tends normally to measure energy consumption in terms of the per head consumption in the population, and a low energy consumption per head is a pretty sure sign of how poor the country is. The higher the wealth, the greater the consumption. Thus an American uses up *a thousand times* as much energy in a year as, say, the poorest Indians, and it is precisely to change that pattern of poverty that the poor countries need greater amounts of energy. The second point is that even if their per head energy consumption is only something like a thousandth of that in the United States, many of the poor countries have very large populations. In other words, they need to multiply their sources of energy very many times in order to make a real change possible. One way of doing so is of course to create as many sources of energy at home as possible: from developing coal fields if they have any, from hydroelectric dams if they have rivers, and so on. But even that takes energy to do in the first place, and in consequence many a beneficial scheme gets stuck in a bottleneck of energy shortage. So it is difficult to begin unless you have already begun. But not only that: even if the government of a poor country does achieve the miracle and break through to the point where it has enough additional energy to start making real changes possible, this of itself multiplies the demand for energy. If, for instance, one finds or creates enough energy to industrialise a region or a country where until now the people have lived in grinding peasant poverty, what happens? Industry means towns and transport. Towns mean energy for industry, but also energy for street lighting, energy for heating and cooking at home (peasants can no longer go into the woods to gather fuel or burn the dung of animals for cooking) and so on. And transport obviously demands energy in large quantities.

The result is that even when countries are beginning to achieve some economic development, they run the risk of getting stuck in acute energy shortages, which at their worst might threaten to undo the good that has already been done. In order to keep going, for example, they might have to resort to forms of energy which are so expensive that they take up all the extra wealth which the process of development has begun to create. And here, too, the obvious answer is oil. So in general terms, it is true to say that whatever other sources of energy are available, a development programme is in the end likely to create a very heavy demand for oil.

And even that is not the end of the story. For obviously, the first need of any poor country with a large population is to be able to feed itself. That means replacing traditional agricultural methods by more intensive farming, based on fertilisers, and the control of pests based on pesticides. Both of these are made from oil: fertilisers are all petro-chemical products; many pesticides are. So the demand for oil for industrial development goes along with a demand for oil products for food.

Any programme of industrialisation today, in any of the poorer countries of the world, represents a high degree of intelligence and no small degree of courage in trying to overcome the twin problems of food and energy. But the solution of the question of energy, and in great measure the solution of the problem of food, depends on oil. And this is where the low energy use, which is a symptom of poverty in the first place, comes back in to weaken the hand of the poorer countries in trying to import oil or oil products. Even those with very large populations are in a weak bargaining position when they are dealing with either the international oil companies or with the states that export oil, because they are poor customers compared with the rich industrialised countries whose use of energy is so much higher. The main purpose of the oil-exporting states being to get as good a price for their oil as they can, and the main purpose of the oil companies being to make as big a profit as possible, they both have a greater interest in exporting to the rich countries than to the poor ones. The rich buy more oil, and they can afford to pay higher prices for doing so. The poor buy less, and can't afford the higher prices anyway. So the interest of the oil states in exporting to the rich survives even their political disagreements with the rich – a point I shall come back to later in the book; on the whole, their indifference to the poor survives even the political solidarity which most of the poor have shown them most of the time.

But the relations between a developing country and an oil company have been somewhat different. In their search for profits, oil companies have frequently moved in to help in the much-needed establishment of new sources of energy. They have handled the shipping of oil, built pipelines and depots, and even refineries. They have, in other words, greatly helped the economy; they have also made very large profits – simply by charging higher prices than any amount of investment would justify. But when oil was cheap at the source – that is to say, when the companies did not have to pay too much to the oil states for their oil – it was possible to make large and unfair profits without necessarily ruining the poor customer. This is no longer possible. The result is that hidden conflicts between some of the countries concerned and some of the oil companies are now coming into the open – they are no longer covered by the fact that the companies have, after all, made it possible to develop the economy. There is now a more direct clash of interest between the oil company's profits and the development of the economy. Such clashes *could* lead to demands for nationalisation of the companies' activities, and to counter-threats of retaliation by the companies, or even their parent countries; such situations have already occurred in Latin America. Or they could lead to appeals for alternative forms of help from the Soviet Union in its role as oil exporter – but, as in the case of Ceylon, such an appeal might bring with it a cut-off in American aid, or, as in the case of India, a simple refusal by the oil companies to handle Soviet oil in the local refineries. (In a sense, though, India won: it was at least able to force the companies to reduce their profits very substantially.) Or they could lead to attempts by the developing countries to deal directly with the oil-exporting states. Such attempts, however, would imply that the oil-exporting state would have to help its customer to pay for the oil in the first place, or at least it would have to give generous rates on long-term loans to build bigger facilities for handling the oil and putting it to use. One such deal was made in 1974 between India and Iran. But on the whole, the oil states, especially the Arab oil states, have so far shown remarkably little interest in helping in this way. From the vast sums of money at their disposal, they have set aside some very paltry sums indeed for any such programme.

In general, the poor countries have been "fined" some 10,000 million dollars a year because of the rise in the price of oil. The *total* flow of aid to those countries, including aid from the oil producers, is now 7,000 million dollars a year. This does *not* mean that the poor countries are now 3,000 million dollars a year worse off. It means that they have lost all the aid which was going to them before, and which was helping them to develop; on top of that they still have to find an extra 3,000 million dollars from somewhere purely to pay for oil and oil products.

The majority of the developing countries therefore now find themselves really caught between the devil and the deep. The oil companies are not showing themselves at all helpful, and most of the oil states are showing themselves far from generous. This is not true in every case: the

Soviet Union, as an oil exporter, has been ready to offer lower prices where it was possible to refine and handle Soviet oil independently of the companies; and in Latin America, the major oil-exporting country, Venezuela, has attempted to come to an equitable set of arrangements with her sister-states. But on the whole, the rise in oil prices and in the prices of oil products has been disastrous, and will probably continue to be so. The cost of fertilisers, for instance, is now getting on for 300% of what it was in 1973, and India is expected to have to make do with a million tons less. Considering that in 1973 it only used two and a half million tons anyway, this represents a very drastic reduction, and it is estimated that it will mean some ten million tons less grain. What this means in a country which was already experiencing famine on and off before, cannot even be imagined. (And now, it is worth adding, there is much less food available in the world as a whole than there used to be, largely because the Soviet Union has bought up most of the surplus stocks in the United States. World stocks of food are now just about sufficient to last for six weeks above the level of current consumption. So that if India, with ten million tons less grain, experiences another famine, it will be paying the delayed price of Stalin's disastrous agricultural policies, and the current price for the wealth of the Arab oil states.)

One faint hope in this threatening situation is that more oil might be found in the poor countries themselves. Certainly, nearly everyone is looking very hard. To take the Indian example again, after a rather half-hearted exploration by an American oil company which produced no very promising result, the Soviet Union and Romania offered India the technical aid to undertake another search. This time, it was largely successful, and India now produces nearly a third of what she uses. (Not a third of what she *needs*!) Moreover, many countries have offered the oil companies such a handsome profit if they do succeed in finding oil that they can hardly refuse to look – and these terms are also usually tied in with guarantees from the company that if it does find oil, it will actually exploit it – rather than just sit on it to keep out the competition. But so far, though there have been strikes, sometimes of some importance, and particularly in Latin America, the hope that this will do much to solve the problem of oil prices over the next ten years remains pretty slim.

Meanwhile, the arguments go on. Should the poorer countries build more refineries because they haven't the foreign currency they need to buy the oil products? But if they do, who will help them with the money in the first place? The companies are usually unwilling, because the amount of oil that a poor country could afford to buy and then refine would probably not justify the cost: it is more profitable for the company to use the refineries it already has elsewhere and then transport the products to the customer. But if the price of oil products is going to be such that the customers really have to cut back on what they are already buying, how long will that be true? And it is even more complicated than that. The advanced industrialised countries have by now begun to pollute their environments so heavily that they are hesitating to build any more refineries – especially in the United States and Japan. This means, naturally, that the demand for oil products will go up among those who can afford to pay the prices they command, particularly for fuel. Which means in turn that a competitive rise in prices for oil products might take place, over and above the rise which results from increasing the price of oil. So in the end, it might be worthwhile after all for the companies to help with the building of new refineries in other parts of the world. What does seem certain is that if they are not built, the poorer countries now face the terrible prospect of getting poorer and hungrier in spite of all their efforts. Even if they are built, the cost of oil is already such that they will probably only be able to keep things ticking over.

Among the developing nations, one country is an exception to this threatening network of complexities. That is China. The reason is that China is determined to remain self-sufficient in its fuel requirements, though, unlike the Soviet Union, it did not start out completely with the intention of doing so. But precisely as a result of its conflicts with the Soviet Union, from which it hoped at first to be able to go on importing, it has become an isolationist country in terms of energy. In this it is helped by the fact that it has large resources of coal, considerable resources of oil, and even more by the fact that such a large proportion of China's population still lives on (and off) the land – some 80%. Given the fertility of so much of China, where the heaviest parts of the population are concentrated; the order that the Chinese have brought into their farming system; and the decisions of the Chinese leaders *not* to drive peasants off the land or starve them while they supply the towns, as Stalin did, this is a perfectly workable situation. Indeed, the first priority of the Chinese revolution has always been to improve the material conditions of the peasants and promote the growth of a rational agriculture. This means that the energy consumption per head of the population remains low, because the multiple demands involved in industrial development have only happened in a few areas; and that the soil, especially when fed by the natural fertiliser of people living on the land, is still adequate to feed the population. But all the same, it also means that China's industrialisation is comparatively slow, and that deliberate restrictions have to be made on the uses of energy. For some years, for instance, it was difficult for Chinese pilots to train properly because there was not enough aviation spirit in the country. This is no longer true but the restrictions on available energy still apply. The only question is how long China will continue to find it acceptable. If, for instance, it wishes to trade with other countries, it might need to use more energy to make use of the products that they are able to give. But the leadership is probably confident that, for as far ahead as they need worry, the expansion in the uses of energy can be satisfied from new discoveries of oil, either in the country itself or in the seas around it.

But with the exception of China – and, just possibly, Argentina, Chile and Brazil – the developing countries of the world are caught in a series of terrible problems. In facing them, it is clear that to some extent they have the same interests as the other countries which rely on importing oil, particularly in the prospects of increasing supplies and – if possible – reducing prices. But whereas the major oil companies have on the whole served the interests of the advanced industrial importers, by cushioning them against the threat of a sudden cut-off in supplies through the wide variety of sources and the flexibility of transport that they have at their command, they are not on the whole serving the interests of the developing nations, because the one thing that they cannot cushion *them* against is a catastrophic rise in prices. Indeed, the oil companies have themselves helped to keep prices high over a very long period, and especially in the areas which can least afford them. More than that: in the end, the profits of the oil companies have been the profits of the Western economic system as well; the others have not shared in these profits. And the technology which the oil companies have helped to develop has formed part of the economic miracle of the Western world, while it has only marginally helped in the direct development of the poor. So the interests of the rich importers and the poor importers are by no means identical.

What is true for both is that the rise in oil prices is going to force them to re-think all their earlier assumptions about the future economy of the world. I called this first section of the book by the words of a member of the Dutch government at the end of 1973: "It will never be the same again". The idea that cheap energy could go on being available for ever has suddenly disappeared, and with it have disappeared the cocksure expectations of the rich and many of the desperate hopes of the poor. In fact, it should have sunk in some time ago, at least from 1970 onwards, that energy was now going to be expensive instead of cheap, but while some people did notice this threat, most chose to ignore it. How it came about, and what its causes were, will be the concern of the next part of the book. But in order to understand that, it is necessary to look first at the major questions: who has the oil, and how the oil companies work.

PART TWO

HOW IT WORKS

3

WHO HAS THE OIL

The question of where oil is found is one of the most important in understanding the whole nature of the oil industry, and of the importance of oil in the modern world. For it is this which makes the question of oil a *political* question as well as an economic one.

The reason is this: if the availability of oil were measured against the world's need for oil, there would be plenty for the immediate future, and perhaps for as long as it is needed (depending on the size of the fields now being explored in Alaska, Eastern Siberia, the North Sea and so on). In any case, the reserves of oil that are already known to exist in the world are ample for the time being. From which it should be possible to argue that with plenty of oil around, there is no need to keep prices at their present very high level, and it is a short step from that to argue that indeed prices are bound to drop in the long term. A balance, so one might say, is eventually bound to be struck between the fact that it is available and the fact that it is needed; and this balance will be reflected in more reasonable prices for oil that is available in greater quantities than at present – the classical notion of "equilibrium" of the economists. This is indeed the view that has been taken by one of the most distinguished American experts on the oil industry, who, from a knowledge of its workings and costs which is second to none in the world, presents a highly convincing case that there is no need for undue concern over either shortages or prices in the long run. But this expert, Professor Adelman, argues purely from economic factors. The question still remains as to whether these economic factors are going to be decisive, or whether they will come very heavily under the influence of political factors. A great many other experts on the oil industry tend to believe that the political factors are likely to prove decisive for a long time to come.

The political aspect of the whole question of oil arises from the manner in which two considerations affect each other. The first is the history of the workings of the big international oil companies, which, not to put too fine a point on it, have shown great ability in exploiting in their own interests first the countries from which they drew the oil and then the countries to which they sold it. The second consideration is: who possesses the biggest reserves of oil available now, and how are the states which possess it going to react – both to the companies and to the countries which need to import oil? In this chapter, I shall look at this second consideration first. Who has the oil?

In doing so, I shall discuss only known reserves of oil. There is very little point in speculating about how much oil is going to be available from unknown reserves, like those in the North Sea. Judging from the history of oil exploration, the new fields there and elsewhere might turn out to be very much bigger than anyone at present knows, and I have already mentioned in the previous chapter that while we *know* that the North Sea could supply a high proportion of Europe's needs for the next generation, it *might* be true that it could do so for a hundred years. But people can't base energy policies on what might be possible, and I shall concentrate on what is known about oil reserves at present.

One other point follows from that: the period over which the present picture of the world's oil reserves is likely to be accurate is perhaps not more than for the next ten years. Thereafter, not only will new fields of production be in operation, but more will actually be known about where other oil reserves are and how big they are likely to be. But for ten years or so, the political character of the oil industry, and the decisions that will be made about supplies and prices will be determined by the picture that is still accurate today, of who has what reserves.

Probably the best figures to take are those of 1972, since these provide a fairly full guide to what the world oil situation looked like in terms of *proven* reserves before the complications of new strikes arose – which, as I have suggested, are going to mean more oil in the long run but are not likely to affect things very much before the end of this decade. The picture given by these figures is dramatically clear. The total known reserves of crude oil in the world then stood at nearly 100,000 million tons. Now of this, the United States – the world's biggest producer – actually had only just over 6%. In other words, there is a gap between being the actual producer of oil and having oil in reserve under the ground. In the American case, this gap has been steadily growing for much of the century, and has now reached a very high level. Equally, the Caribbean countries, which include Venezuela, once one of the world's biggest producers, now possess under 3% of the world's reserves. Western Europe only has just under 2%. Most of the rest of the first 50% of reserves in the world is shared roughly equally between Africa with just over 15%, and the Soviet Union, Eastern Europe and China, which, taken together account for almost another 15%. And where does the second half of the world's reserves come from? It is in the Middle East. The Middle Eastern countries, taken together, possess over half the known reserves of oil in the world: 53%.

Now, the first thing to note about this startling picture is that it represents a very great historical change. For until the Second World War, not only was most of the oil produced outside the Communist countries, produced in the United States, Mexico and Venezuela, but it was generally expected that this could go on being the pattern. Oil did of course reach Europe from the Middle East long before that, notably from Iran, but the bulk of West European dependence on oil was met from the Western Hemisphere. But the switch to an oil-refining policy that I mentioned in the last chapter, coupled with the growing American need to import oil after 1948, meant that the Middle Eastern sources had to be explored and exploited more thoroughly. In other words, *as the dependence of the world on oil has increased, so has the importance of the Middle East grown with it.* The picture might be modified slightly during the ten-year period that it is realistic to consider – for example, Japan might be able to meet more of its needs from Indonesian oil; but on the whole this kind of modification will make little difference. The growing dependence on oil and the growing importance of the Middle East will continue to go hand in hand. Even the United States is importing more and more from the Middle East, though Venezuela remains its chief supplier.

The second point to notice is that, in transport terms, the Middle East is conveniently near one of its main customers: Western Europe. In the sense that for years tankers have not been able to use the Suez Canal, and that now, even if it is re-opened, many of them will be too big to do so, the Middle East is further away than it was; but that has been largely made up for by a network of other arrangements, including pipelines to the Mediterranean. These in general make Middle Eastern oil physically the easiest to get at for European countries. In fact, they now have a vested interest in oil from that region because such an extensive set of arrangements already exists. Japan is of course in an opposite position: Middle Eastern oil has to travel huge distances to get there, but for the time being Japan has very little choice. But the fact remains that for one major customer – Western Europe –

alternative sources of oil would not necessarily be very tempting unless, of course, they were under its own control.

The third point worth noting is that at least some Western European countries – most obviously, of course, Britain and the Netherlands – make very considerable profits through the operations of their oil companies in the Middle East. While they continue to pay high prices for the oil, and incur severe deficits on their balance of payments as a result, some of the loss is compensated for, overall, by the fact that the companies bring home a lot of money. This goes even more for the United States; and for this reason governments have in the past encouraged companies to seek relatively high oil prices because they thought this was an overall benefit to their own economies. How wrong they were, they have just begun to discover, but in the meantime the governments have helped to build up their own reliance on oil in the Middle East, a region where it could be obtained most cheaply but sold for a good price outside.

This pattern of increased reliance itself helped to stimulate exploration of the Middle Eastern oil deposits even further, so it is not altogether surprising that the Middle East now has such gigantic known reserves. But the most important consideration is that in that area the oil is easy to get at and cheap to lift out of the ground. I have already indicated the tremendous disparity between the real cost of a barrel of American oil and a barrel of Saudi oil, and much the same goes for Kuwait and Abu Dhabi, two other leading producers on the Persian Gulf. This was, above all, the reason for greater and greater dependence on the oil from the Arabian sands. What it led to in the end was the realisation by the oil states that they could themselves make vastly bigger profits from the sale of their oil than they were getting. In consequence it led to the complicated relations which have arisen today between the oil states themselves, the companies, and the oil-importing states. These relations will be explored better when I have considered how the companies work; but the point at present is that Middle Eastern oil is *essentially* cheap, and that the prices which have now grown so enormously have nothing to do with the cost of producing it. The prices reflect, first the fact that a large part of the world has grown to depend on the Middle East for its oil; and second the way the oil states have reacted both to the countries that buy it and to the companies themselves.

In other words, they reflect the politics. Now, for all their internal differences, ranging from the autocratic form of modernisation which has been undertaken by the Shah of Iran, through the intensely traditional Islamic society of the Wahabi kings in Saudi Arabia, or from the paternal social welfare of Kuwait to the fervent mixture of Islam and socialism which characterises the revolutionary regime in Libya, the politics of oil have very little to do with the political relations of the oil states. The politics are really composed of three elements. The first is the one I have already mentioned – the relations between the oil states and the oil companies. This used also to mean the relations between the oil states and some of the oil-importing countries too. The United States began to face difficulties in its relations with Venezuela because it backed American oil companies against the government of that country; and its relations with some of the Middle East states have been complicated by its backing for the companies there too. Similarly, Britain faced a major crisis in 1951, when Iran nationalised British commercial oil holdings there. Much of this pattern continued until recently. But by now, it is not automatically assumed that what is in the best interests of the companies is necessarily in the best interests of the countries behind them; and in this sense the companies are left to sort out their own relations with the oil states as far as possible. None the less, this set of relations does of course continue to influence the relations between the oil states and the rest of the world.

The second element in these political relations follows from the division that has grown up in the first. It involves the state-to-state relations between countries that export oil and countries that import it. In the wake of the crisis of 1973, different attempts were made to organise these relations on an international footing. President Nixon called an international conference to do so in Washington in March 1974; but it was notable only for the fact that the European countries politely refused to co-operate. (In the case of France, impolitely.) The European Economic Community has also tried to organise its own set of relations, with the Arab states in particular, on much the same basis, and here the Arab countries showed more interest – to the point of gate-crashing what was supposed to be a private EEC conference in Copenhagen. The "Euro-Arab dialogue", though it does not yet exist, is the outcome of a series of meetings of this nature; but not only does the dialogue not yet exist; there is not yet the basic agreement among the West Europeans as to what it should be about. International reactions to the crisis of 1973 will be considered in a later chapter. But it is now clear that on whatever basis they operate, the politics of oil will be conducted at least as much through the relations between the exporting and importing *states*, rather than through the relations between the exporting states and the *companies*. Whether it is an unmixed blessing to do endless business with Sheikh Yamani, Saudi Arabia's tireless Petroleum Minister, is an open question; but at least he gives the facts rather more openly than the companies when one is trying to come to an arrangement with him.

The third element is one that matters from time to time, but on the whole does not matter nearly as much as headlines in the press would seem to suggest. This does not concern the way that oil questions have begun to affect political relations but the way in which political relations in other fields have sometimes affected oil. The central point at issue here is of course, Israel. On two occasions, in 1967 and in 1973, the Arab states have cut off supplies of oil to those whom they regarded as Israel's friends (not many people in Israel would share that opinion of a number of them) or threatened to do so. More than that, the king of Saudi Arabia has on occasion emerged from his harem to announce that he wanted to pray in Jerusalem, and that unless Jerusalem were handed over to the Arabs, America would find itself running short of oil. But, although the Soviet Union has encouraged the use of the "oil weapon" in this way, the threats have never really stuck. They have had remarkably little effect on the progress of peace negotiations between Israel and the Arab states. Threats and embargoes have been quietly lifted after a time, even on the two countries which the Arab states had decided were Israel's two chief supporters – the Netherlands and the United States. Even when they were in operation, they were not fully accepted by all the countries concerned: the Shah of Iran, who of course rules a country which is not Arab, treated them with scarcely disguised contempt; Iraq and Algeria decided that it was against their own interests to observe them; and in the end even Saudi Arabia came to the same conclusion. So, although one might be tempted to suppose that the Arab-Israeli conflict has played a large part in determining the relations between oil-producers and oil-consumers, it is really not true. At most, the war between the two sides encouraged the oil states to raise their prices even faster in 1973 than they might otherwise have done; and in general Israel impinges on the politics of oil only from time to time and only to a limited extent.

One other question of politics at large is connected with the politics of oil, and that concerns the relations between the countries of Black Africa and the white government ın South Africa. What happened here was that after the war between Israel and the Arab countries in 1973, the Black African states, which had all been persuaded or pressured by the Arab countries, and especially Libya, into breaking off their relations with Israel, demanded that in return for this show of solidarity the Arab countries should suspend oil exports to South Africa. They also hoped for some help from the Arabs in paying the new prices for oil, which were by then climbing out of their reach. In the event they got a scant reward on either count. Iran, once more, decided that business was business, and made it clear that it would continue to trade with South Africa provided the price was right – and the argument that if one does it, why should another not share in the profits, also appealed to the oil companies. And as for help: well, Saudi Arabia earns £20 million *a day* from selling oil, but all that the Arab countries between them managed to set aside for help to the African countries in buying oil was £180 million *a year.* This is from the pooled wealth of the Arab oil states, and it has to be shared out between the pooled poverties of the Black African countries. So much for solidarity. To be as fair as possible: the oil states have shown more interest in helping Islamic countries generally, and by the summer of 1974, some energetic discussions were already taking place on the basis of religious solidarity. So far they have not led to anything that amounts to much; but if they do, it will certainly doubtless show that it is not politics which has entered the field of oil relations, but religion.

Finally, it is worth remarking that embargoes for specific political ends are in any case very hard to sustain. Oil companies have more than one trick up their sleeves when it comes to matters of that kind. They buy oil in Nigeria, or Indonesia, or Venezuela, or Iran, as well as in the Arab countries. They have fleets of tankers that go all over the globe. All they have to promise is that they will buy oil from one particular country at a particular price. They then sell it to another at a particular price. How can one always know where the oil ends up or where it comes from anyway? To a degree, one can, by the quality of the oil when it is delivered. But on the whole, it is very difficult to keep a check on such a wide range of activities; and when the Netherlands was subject to an official boycott, they only went short of oil to the extent – not that the boycott bit – that tankers from the oil companies steamed up and down outside Rotterdam, waiting for prices to rise. In the end, the only way for oil-producing states to make sure that none of their oil reaches a country they dislike is for them to cease production altogether; and while they may be prepared to do this for a short time, there comes a point when they have to consider the relation between cutting off their nose and spiting their face.

Altogether, then, what I am arguing is that politics outside the oil business itself do not count for very much in the workings of that business. But I should add that so far the real contest of wills has not happened. If, for example, the Western European countries, or Japan, had stubbornly proclaimed their support for Israel during and after the 1973 war, and if the Arab states had stopped all production of oil immediately, it would have been illuminating to see who held out the longest. But all that has happened so far is that after a symbolic skirmish or two, things went pretty well back to normal. Oil barely affected the outside politics, and the outside politics scarcely had any impact on the oil. What was *not* normal was the change that was occurring in the political relations between the various partners in the oil industry.

To understand these further, it is necessary to consider rather more closely the question of who has the reserves and how this affects the prices. To put it at its crudest, Saudi Arabia could declare tomorrow that the price of oil should henceforth be half a dollar a barrel – and everyone else in the Middle East would soon fall into line. This is for three reasons. First, the real cost, as I have already suggested, of getting a barrel of oil out of the ground in Saudi Arabia is ten cents a barrel. Thereafter, everything represents a profit for somebody. If the Saudi price were half a dollar, it would mean a profit of forty cents a barrel for the Saudi government. After that, transport costs, oil company profits and importing government taxes would have to be considered before the price that the consumer actually pays could be arrived at. But none of the other oil-exporting countries would have to consider these things. All they would see was that Saudi oil was going to the oil companies at fifty cents a barrel. Now if Saudi Arabia had very small oil reserves, they could afford to wait until those reserves were exhausted, and then charge very much higher prices when the companies came back to them, cap in hand. But Saudi Arabia does not have very small reserves – it has the largest known deposits of any country in the world. Which means that other countries would have to wait a very long time indeed before they could begin to sell their oil again in competition with the Saudis. And that is the second reason: those vast reserves in Saudi Arabia mean that nobody else, in the present state of world demand for oil, could afford to suspend production for years on end, while the Saudi government continued to make vast profits. They would go bankrupt. And this indicates the third reason: most of the other oil-producing states have very little else to offer. (I am still speaking about the Middle Eastern countries with their half of the world's known reserves.) Some of them do, of course: Iran, for instance, is turning into a very diverse society – though largely on the basis of the capital it has made out of oil. But the case of Libya is more typical. Before the discovery of oil there, Libya's largest export was scrap-metal, culled from the immense amounts of armour left lying around from the Second World War. Kuwait and Abu Dhabi have nothing else to offer at all. Oil is their only source of income and they have to sell it now, at as good a price as they can get, before other deposits are discovered, or before the world begins to turn to other forms of energy. Which is why they could not afford to compete with the Saudis. Fortunately (for them) the Saudis are in much the same boat, and have no interest whatever in reducing the price of oil by any substantial amount, and so the question is not likely to arise.

But what this means is that in a situation where many of the most important oil-producing countries have very little but oil to fall back on; where they have the greater part of the world's oil deposits to make the best use of that they can (and here it is worth remembering that they not only have such a high proportion of the reserves, but an even higher proportion of what is available for export – America's six per cent doesn't represent any exportable oil at all, considering that it is an importing country anyway); and where one of them is dominant in determining the price of oil; then the *pace* at which that one decides to exploit and sell its oil determines what happens everywhere else.

This question of pace is very important. Suppose that Saudi Arabia decided to make as much money as possible while the going was good. It would then decide, or so one might suppose, to sell as much of its oil for as high a price as it could get as soon as possible. But the result would not be as advantageous as one might expect. For, because of the higher prices that are being paid for basic materials of every kind – not just oil – the money that the Western world can offer is likely to go on losing value for years to come. In other words, money in the bank is a wasting asset. On the other hand, oil is not: demand for oil is going to remain high for the foreseeable future. Oil in the ground is therefore a far better asset to possess than money in the bank – because if it is sold off gradually it will go on representing a *real* value which can always be shown in higher prices in years to come – prices which reflect what money was worth *then* rather than what money is worth *now*. It is therefore in the Arab countries' interest to slow down the pace at which oil is exploited and sold, to make sure that they continue to enjoy real wealth from oil in the future, rather than watch the money into which they have converted oil slowly become worthless in Western currencies. Hence, they have to try to do two things at once. First, to raise the price of oil as high as possible now; second, to restrict the pace at which oil is taken out of the ground and sold abroad.

In this way, the question of pace becomes linked to the question of price. But it is slightly more complicated, even, than that. Because if the price is raised too high, it can set off such an inflationary spiral in the Western world that the money that an Arab, or other oil state, has already earned is in great danger of losing its value anyway. In this sense, it becomes an interest of the oil state, not only to slow down the pace at which oil is sold, but also to make sure that prices do not rise too far. In some cases, there might even be an interest in reducing prices – and this might shortly start to happen, even if not by any considerable amount.

But at this point, the different interests of the oil states begin to show just how much they diverge. If you are the king of Saudi Arabia, you rule over a small population in a large country which consists mainly of desert. You want to bring about some social changes; but not many at a time, otherwise your throne might be in danger. You also want to show some power in the world – to represent your version of Islamic values in Arab society, to show the Western powers that they can no longer push the Arabs around, and put pressure on them to force the Israelis back. For all these reasons, both internal and external, your chief interest is in acquiring money of real worth and therefore regulating the pace of oil production. This gives you two means of pressure. In the first place you can hope for a long time to come to be able to influence Western policies by the regulation of the oil supply; and in the second place, you can use your money, either in the Arab world, or through investments in the Western world, to favour your friends and put your enemies at a disadvantage. In view of both your traditional sense of values and of the nature of your kingdom, you are above all not interested in pushing prices too high for the sake of development at home. For these kinds of reason, Saudi Arabia, whose attitude to the Middle East conflict is a good deal more unbending than that of most actual participants, and whose approach to the secular West is far from approving, has tended to emerge in practice as a preacher of relatively "moderate" policies. Suppose, though, that you are the Shah of Iran. Then your list of priorities is completely different. You rule over a relatively large population in a country which is important for other reasons than oil alone: for one thing you have a common border with the Soviet Union and you also have memories of past grandeur which are far from dead. You also feel that unless the country is developed very fast, you will be seen, rightly or wrongly, as a reactionary autocrat who stands in the way of progress, and your throne will be in danger from the usual combination of students, socialist intellectuals, and workers who are by now becoming educated enough to know what they want. For all these reasons, you wish to turn Iran into a strong power and a developed country, and to do so very fast. The fact that you are internally repressive (and that Iran holds the world record for executions outside South Africa) does not prevent you from being "modern" or from having high ambitions. In this case, and because you can hope that Iran will continue to develop other forms of wealth than oil alone, you will want as much money as you can get for the oil, and you will want to have it as soon as possible. So your own interest lies in pushing oil prices as high as they can possibly go, and in using them as the base to obtain nuclear reactors, or defence equipment, or whatever, from the Western powers. For these kinds of reason, Iran, which has in the past tended to emphasise its friendship with the West, and which takes little account of the questions which so agitate the Saudis, has emerged as a power which is a good deal more "extreme" in the policy of oil pricing.

In fact both are merely following their own interests, and in this as in other matters it is pointless to classify the oil states as moderate or extreme. The real point of difference between them is whether they are rich or poor. There is, in this sense, a fairly sharp difference between those states whose population is small and whose revenues are large, and those whose revenues are certainly large but who also have a large population and have, or have had until recently, the same problems of desperate poverty as any other country in the developing world. The rich countries are Saudi Arabia and the small sheikhdoms on the Persian Gulf. Here the income per head is the highest in the world (Kuwait actually has the largest income per head in the world) and the only problems are in trying to think of what to do next with the money. The poor countries are Iraq, Algeria, and Iran. All of these are concerned with military power as well as with social transformation — although to different degrees — but they *are* also concerned with social transformation. In no case should one confuse social change with democracy: they are all run by very authoritarian regimes indeed; but none the less great efforts are being made to bring about a set of changes which will relieve the people from the grinding work for a miserable existence which has been their experience up to now. In between is Libya, whose territory is enormous but whose population is small, and whose oil revenues are comparatively recent. It is also in the hands of an extremely unstable regime, which so far cannot be said to have shown any distinctive policies beyond those of squeezing the Europeans.

There is between rich and poor a certain conflict of interests: not of fundamental approach, but of tactics and timing. For the rich, the main interest is to regulate the flow of oil so that it doesn't all get transformed too fast into money, but continues to provide real revenues for years ahead. In this sense, Libya is clearly among the rich. For the poor, the main interest is to transform as high a proportion of the oil as possible – not only into money but also into other projects of development aid from the West, so that these countries themselves can create more wealth for themselves, and do not have to rely nearly so much on oil in the future as they do now. Obviously, both sets of states have a strong interest in raising the price of oil, though the poor are prepared to go further than the rich. But whereas the richer ones, and Saudi Arabia in particular, have shown that they are willing to suspend their oil production in such conflicts as the 1973 war, the poorer ones have not, since for them it is important to maintain a rhythm of revenues which can be turned into a rhythm of development. Equally, the best way for them to turn the revenues into development is to accept payment in the form of technical projects from the industrialised powers – which means keeping up some form of contact with them. For them, therefore, the "Euro-Arab dialogue" might prove to be of some importance.

Beyond such differences in tactics, they do, however, have a more general interest in common. This concerns the way in which the revenues from the oil states are shared among those Arab states which do not have oil. There is a general agreement on an obligation to do something about this, though the form that it takes might vary. Libya, for instance, has been willing to offer political union with

Egypt, which would have meant in effect that Libya's money could have financed Egypt's economic and social development – but the terms on which the unstable Colonel Ghaddafi insisted ruled this out in the end. Saudi Arabia, on the other hand, has been willing to finance specific projects, again notably in Egypt. The relations between the oil-rich states and the rest have certainly not been defined yet, and they might only achieve a gradual definition through parallel conversations among the Arabs themselves, and also between the Arabs and the industrialised powers of the West. Meanwhile, one might add that even those Arabs most anxious to see the oil revenues being used to change the standards of living in the Middle East in general, and not merely among the rich states, are by now sick to death of Western advice on how to invest their revenues in order to help each other.

So far in this chapter, I have argued that the politics of oil are likely to be an important and lasting element in determining the price of oil, and that one cannot take a purely economic view of either the future price or the future availability of oil. I have also tried to show how questions of price and questions of making the oil available turn on a number of different considerations which vary from one country to the next, and how these have to be harmonised in working out the level both of price and output. I have dwelt at such length on the Middle East, not only because that is where most of the exportable oil reserves are, but also because it is the different considerations of the Middle Eastern states, the manner in which they compromise with each other, and the way in which they present their conclusions to the other powers, which are fundamental in reaching the decisions. But behind this general picture, it is necessary to look a little more closely at the pattern of oil politics that has developed in the relations between individual states and the oil companies and their home markets.

Here, it is important to consider first a country that is not in the Middle East at all, but which first experienced all the contradictory pressures of finding itself a major oil-exporting state: Venezuela.

Venezuela began to develop its oil reserves rapidly during the Second World War – only partly as a result of the war itself. It was also in part because oil companies had been expelled from Mexico – until the 1940s a major producer – and began to look elsewhere for their profits. By the end of that decade, Venezuela was the fastest-growing oil country there was. But this immediately created a set of problems of the kind which have since become familiar. If one Venezuelan government (the Acción Democraticá) of 1948 talked about nationalisation, the oil companies made it very clear that they had lost interest – primarily by keeping production at the level it then was, and ceasing exploration for new wells. But if the dictatorship which shortly followed opened up the country to the companies – as it did – Venezuela soon risked becoming too dependent on oil in order to keep the economy going; and becoming dependent in a way, moreover, which made it subject to the companies' planning and profits. As a result of these different experiences, the government which succeeded the dictatorship (Acción Democraticá again) came gradually to the decision deliberately to restrict the output of Venezuelan oil, in order to reduce both these forms of dependence. The result is that Venezuela, once one of the front runners in world oil production, was easily overtaken by the Middle Eastern countries in the 1960s. (Its known reserves are also of course very much smaller, but one should add that this is partly because the decision to restrict production was taken, and therefore exploration for new reserves has not been anything like as intensive as in the Middle East.) In this case, restrictions on the production of oil have actually helped the Venezuelan economy by encouraging it to develop faster in other ways. But clearly, it has not reduced Venezuelan interest in high oil prices, and that country too has been active in trying to bring about increases in price over the past four years.

Venezuela is in many ways a classic case: the case of a weak and rather poor country trying to frame a set of policies which is in its own interests rather than those of the huge companies which threatened to dominate its economy. But it was able to succeed partly because in the later stages it had the unexpected backing of the American government – concerned by then as the United States was with the "threat" to the Western Hemisphere from Castro's Cuba, and anxious as it had become to back the more stable governments in the area. This helped to bring the companies under better control but it has not helped Venezuela in its further dealings with the United States – where it has been anxious to obtain special concessions in the American market and is in some degree entitled to enjoy them through the treaties of association between the United States and Latin America. The only concession it has managed to get is for the export of oil-based feedstuffs to Puerto Rico – Venezuelan oil, as it happens, being particularly suitable for cattle cake.

What the whole case demonstrates is that oil-producing countries have had to frame their policies around a conflict of interest with the oil companies, and that it has been possible for them to win usually only when the home governments of these companies are either too weak to protest – which has been increasingly the case since 1970 – or else have other reasons for supporting the oil states against their own companies. The lessons have not been lost in the Middle East. Oil companies and their activities have brought immense benefits to many countries but these activities have been undertaken for gain to the company, not for benefit to the country, and where the two have clashed, the company has generally shown that the arguments for gain win. From the viewpoint of the countries, this looks very like imperialist exploitation, and when they are suddenly given the ability to reverse the balance, it is thoroughly understandable that they should seize it with both hands. And, to take it further, it is even then not the profits of the oil companies which suffer, but if it proves to be the consumers and taxpayers in the countries which import oil, well, that's up to them. Perhaps an Arab might be inclined to think that it is time the oil companies' customers stopped blaming the producers for everything and cut the companies down to size from the customer end, as they themselves have been trying to do from the producer end.

For them in the end, the politics of oil is part of a revolt against imperialism, and an attempt to regain some of the profits they have lost from the use of their resources in the past – in some cases, their only resource. And this is the final political framework, embracing both the questions of inter-Middle Eastern politics and the question beyond that of the politics of dealing with the advanced industrialised countries over terms and payment. It is this, finally, which makes it almost irrelevant whether a particular government in a particular oil state is "conservative" or "revolutionary", "pro-Western" or "pro-Soviet". They are all concerned to re-make the terms on which they offer their oil for sale to the rest of the world and variations within that pattern represent fairly clear differences in national interest, not in the outlook of the governments. Nowhere has this been illustrated more clearly than in the other major clash of interests between an oil state and the outside companies and government, the case of Iran.

Iran is an empire which tries to combine conservatism in government with radical change in the society. What the Shah calls, with perhaps more accuracy than he intended, a "White Revolution". But it was this conservative and authoritarian government, a "traditional friend of the West" which in 1966, threw down a particularly difficult challenge to the oil companies. Iran was trying to establish a new principle in oil production: namely, that increases in the production of oil should be related to the development requirements of the country from which the oil came. Now at this point Iran had a highly successful and expanding oil industry, growing at the rate of 15% a year. But it argued that this was not enough: the needs of its population demanded more than that; and it was also able to point out that exploration of more fields in the country had been successful enough to warrant an immediate increase in oil output. The companies could hardly argue with this – they had done the exploration themselves, but (this was in the middle of a price war in which companies, especially petrol companies, were trying to undercut each other) they knew that if they did increase their output, either they would have to reduce it somewhere else in the Middle East, or profits would drop. In addition, though Iran has very conveniently located oil fields, the new output would in any case be less profitable than the oil they were already extracting from other areas in the Middle East. In consequence, they fought the Iranian demands as hard as they could. But in the end they had to give way and in doing so, they helped to finance further economic and social development in Iran.

The difficulty was, though, that their surrender was not in the best interests of the other Middle Eastern countries – which, at least for a time, stood to lose by Iran's gain. But the lesson that has gradually been drawn from this and other episodes is that if the oil-producing countries stick together, they can work out arrangements which while they are inevitably compromises, are none the less in the general interest of all of them. This is the logic behind the increased activities of the Organisation of Petroleum-Exporting Countries (OPEC) which was founded in 1960, but began to acquire teeth in the late 1960s, and behind OAPEC, too, the more recently founded Organisation of Arab Petroleum Exporting Countries. More will be said of these when I have considered how the oil companies actually work.

Before going that far, it is worth looking in a little more detail at which countries actually produce most of the oil in the Middle East's share of more than half the world's reserves – and such a very much higher proportion of the world's exports. There are four really big oil exporters there: Iran, Iraq, Saudi Arabia and Kuwait. Two of them, Iraq and Iran, have old oil industries whose fortunes since the Second World War have varied enormously for political reasons. (In itself this is a strong argument for considering the political reasons on at least the same footing as the economic reasons when trying to judge the levels of output and prices in the future.) The other two have relatively new industries which have shown a very steady rate of growth, and have not been subject to change because the political situation has always remained stable. Iraq has also experienced peculiar problems with transporting its oil, because the first pipeline ran from there to the port of Haifa in Palestine; then when the State of Israel was established, the Iraqi government refused to use this any longer. A new one was constructed through Syria, but this has the unique distinction of being vulnerable to both Israeli attack in the event of war (especially the refineries which it feeds) and to Syrian or Palestinian attack in the event of a peace that they object to. Moreover, its sea outlets are complicated by a dispute with Iran over the control of a vital Persian Gulf outlet, the Shatt-el-Arab. Because it has experienced such vulnerability and such variable fortunes, Iraq has shown a strong interest in keeping oil production up and keeping it regular, even through the international disputes of recent years. Its changeable relations with its neighbours have also inclined it to rely heavily on the Soviet Union, not only for defence but also for technical assistance from time to time – and yet the Soviet Union is the power which has most consistently advised the Arab states to use the "oil weapon" against Israel – i.e. to shut off the oil. All in all, it is a complex situation, and one made even more complicated by the fact that Kurdish nationalists in Iraq operate very near the main oilfields, and could even attack them. But what needs, again, to be emphasised is that Iraq can only deal with these complications through agreement with its fellow Arab states.

Iran, not an Arab state, has also experienced fluctuations of fortune, although it has the oldest developed oil industry in the Middle East. Once again, the causes are political. The chief setback it experienced was in 1951, when Mussadiq (the Prime Minister who used to give press conferences in his pyjamas, weeping the while) nationalised the British oil holdings there. He also forced the Shah to flee. But the results were disastrous. At that time, the prospects for great increases in the Iranian oil production were very favourable – indeed, output had already doubled between 1945 and 1950. But the Anglo-Iranian Oil Company (as it then was) had negotiated agreements with the Iranian government which made the oil the property of the company once it had been taken out of the ground and paid for. In other words, Iran had no right in international law to sell its own oil anywhere. And the company threatened legal action against anyone who bought it: to such effect that nobody did. Meanwhile, all that happened was that oil-fields were developed in Kuwait to make up for the lack of oil from Iran, and the country soon faced bankruptcy. In the end, a combination of internal and external conspiracies brought the Shah back, disposed of Mussadiq, then set the stage for re-negotiation between the country and the company. The results of this re-negotiation were that oil was no longer the property of the company, which in law, now became only the agent of the Iranian government in extracting the oil, and were thereby entitled only to a (still very heavy) share in the profits. And it was this new set of legal facts which enabled the Shah to stand up to the companies when he demanded a higher level of output from them in 1966. By then, he was free to sell his oil to anyone he liked, including the Russians, if the companies didn't like it. (The Russians of course didn't want it for themselves. They would merely have done the Shah's marketing for him, and so confronted the companies with a total loss.) This early history of Iran's dealings with the companies goes a long way to explain why it has been in the forefront of plans to increase prices, and also why, for so long, it has seen these increases in profits from oil as a way of getting on with social development.

The two other giant producers in the Middle East are Kuwait and Saudi Arabia. There is really not much to say about them. Since the late 1940s, production in both has gone ahead in very steady leaps and bounds. But there is one difference between Saudi Arabia and the rest which is rather important. It is the only country which deals with American companies alone (partly because Shell, after some initial exploration, decided that it was not worth developing. It must be one of the most expensive mistakes in history). This means that, while all the oil states by now deal with a set of companies, and never with one company alone, Saudi Arabia deals only with American companies and the American government behind them. It does at least make it easier to relate the politics of oil to oil *in* politics; but it also means that Saudi Arabia is quite exceptionally important in political and diplomatic questions for more reasons than the simple fact that it possesses the largest oil reserves. On the other hand, it does not necessarily mean that the interests of the country and the interests of the companies are likely to be identical. Any outsider, looking at the Saudi system of government and society is bound to wonder how long it will last. If the outsider is an oil company – or the group of American companies which trades there under the name of Aramco – it is likely to suspect that one day there will be a revolution, possibly a bloody one, and that when it happens the oil will be nationalised. For the companies, therefore, the chief interest lies in getting as much oil as possible out of Saudi Arabia as fast as possible before it could all be nationalised, and this means that they would like a very high pace of production – even accompanied by high prices. But since the Saudi interest lies in restricting the pace of production in order to save up revenues for the future, there is a direct clash of interests. The State Department, committed to working as far with the Saudi government as possible, tends not to put on much pressure here; so in the end, the two governments are working more or less in harmony, with the oil companies in the middle. Not a happy position for an oil company, but none the less they are not doing too badly.

It should be evident from all this that the political interests of the principal oil states are quite different when one compares what one country desires or needs with the requirements of another country. It should also be evident that they have found that the best way of resolving their differences is by working together in dealing with the oil companies. In other words, the organisation of oil-exporting countries, and especially the Arab ones, is not so much an alliance, sharing common objectives, as a trade union protecting certain common interests. A trade union depends on a common employers' system to function properly and it must also have the flexibility to cope with new claims for membership. The first condition was met by the international structure of the oil companies; the second by the fact that most of the new oil fields discovered since the Middle East came to assume its dominant position in the world's oil economy were also in Arab countries – notably in Libya and Algeria. Indeed, Libya is today the world's third exporter of oil, and since 1970 has been directly responsible for much of the process whereby the economics of the oil situation were transformed by more radical politics. But even so, the four main countries that I have mentioned here – Iran, Iraq, Saudi Arabia and Kuwait – have, along with Venezuela, been responsible for 90% of the world's oil exports since 1945. Their voices are also likely to remain decisive for the ten-year period during which one can make any sensible statements about the future. At present, they seem to have a set of interests which are fairly compatible with those of the oil companies – though not necessarily with those of the oil-importing countries. To understand both how this came about, and how the situation is developing, it is now high time that the operations of the oil companies were considered in more detail.

4

THE COMPANIES

Oil companies find oil in the first place; then they take it out of the ground; then they transport it. After that, they often refine it. Then they sell it – sometimes in great bulk, as in the case of fuel oil, to governments or other large concerns; or sometimes they distribute it, down to the smallest quantities to the smallest customer, as in the company-owned petrol stations, which are quite ready to put a single gallon into a single car.

They are in fact concerned with oil and oil products at every stage: production, transport, refinement and development, sales. This means that they are very large and complex organisations. They have to take a great number of factors into account in everything they do. For example, if one is selling oil from Kuwait, compared with oil from Algeria, to Western Europe, one has to balance the fact that Kuwaiti oil means lower costs, in terms of extracting it, against the higher costs of transporting it. But then one has to go beyond that. Some oil is heavy, and full of sulphur, which means that it is hard to turn into a wide range of petrochemical products, and that the sulphur has to be burned off. This in turn means concern with the anti-pollution regulations in the country where the refinery is located. This represents higher costs and probably a lower ability to realise the value of the oil at the end of the process. Some oil is light and contains little sulphur. Less has to be burned off to get rid of the sulphur, there is little need to worry about anti-pollution regulations and the oil can be turned into a wider range of products at the end of the day. These sorts of considerations are vital when one is talking about oil. To give a further instance: it was widely assumed that Israel, after it had captured the Sinai peninsula from Egypt in the war of 1967, would refuse to give it back because it contained a small, but (compared with the size of the Israeli economy) significant oil well. Indeed, it produced more oil in total than Israel needed. But in fact, the oil was a very minor consideration to the Israeli government because it was heavy, sulphurous, and could not be turned into some of the products that Israel needed most – such as petrol or aviation spirit. It would have provided admirable fuel oil – but in the Israeli climate that was not such an important matter as it might have been in Western Europe. So, to go back to the first example: a company will balance the costs of extraction against the costs of transport; but it will then go on to compare these two costs with the other factors involved. Algerian oil is more expensive to extract, but it is light and relatively sulphur-free. In other words, its value can be realised more easily. Kuwaiti oil is heavy and sulphurous, and its value cannot be realised anything like so easily.

These are the initial complexities. But beyond these, there are a host of others. If a company operated *only* in the two countries of Algeria and Kuwait, its task would be relatively simple. But it is likely to be in operations all over the globe. It then has to start calculating costs of transport against costs of extraction, the kind of oil which is going to what kind of destination ("We can get it there quickly and cheaply, but the oil is heavy and the anti-pollution laws are strict and expensive – probably better to take it somewhere else, and spend more money getting a different kind of oil to this customer") and the time-scales over which the value of a particular kind of oil can actually be realised. The complications of this kind of arithmetic are such that it can take a great deal of forward planning, highly-skilled manpower, even computer- time, to work out the most economical and profitable way to move even the various company tankers, as they are required to transport their cargoes of different kinds of oil to different destinations around the globe. And this complexity brings into the open one aspect of the operations of the oil companies which is sometimes overlooked: they can be extremely vulnerable. The loss of a tanker like the *Torrey Canyon* might not only mean the loss of several million tons of valuable oil as well, and the subsequent damages paid to a government for having to clean up the mess. It can also mean that a very delicately balanced set of calculations is thrown out of gear, and involves a tremendous loss *within* the operations of the company. This vulnerability at the edges of the company's work ('vulnerability at the margin', in the economist's jargon) can in fact play an important part in the ability of governments to put pressure on them, as I shall suggest later.

Apart from the complexity of all but the most modest company, the oil industry is for the most part operated on a gigantic scale. Most of the profits from oil are made by a group of seven companies – the "seven sisters" as they are known by their enemies. Together with the CFP (Compagnie Française des Pétroles) these seven were responsible, between 1960 and 1965, for 63% of all oil production outside the Communist world. Also for 60% of the refinement and for 55% of the distribution and supplies. And if one looks at the specific areas, the figures are just as startling: these same companies were responsible for the exploitation of 86% of the Middle Eastern fields, for 90% of the Venezuelan fields, and for 58% of the Libyan field.

This gives some indication of the enormous scale, power and wealth of the principal oil companies. The best way to an understanding of how they have operated, and, more important, to an understanding of the great changes that have come about within the past four years, when the oil-producing states began to change the rules of the game, is to look at the history of these companies.

Up to now in this book, I have often called these companies international. They are, in the sense that they operate internationally; but the main base for five out of the seven is in the United States. And this indicates the importance of the distinction I indicated when I was discussing the United States: that the companies established international *markets* before they established international *operations.* For most of their history, they have been expanding their operations in order to keep pace with the growth in the demand for oil after the time that the United States was ceasing to be the world's main supplier of oil. This is a pattern that was also influenced by the early history of the companies' operations in the United States itself.

The largest company in the world (not only in the oil business) is Standard Oil, which used to trade under the name of Esso (Esso is another way of spelling SO – i.e. Standard Oil) but which now goes by the name of Exxon. Report has it that the scale of operations of this company is so large that it ventured into one, or even two, languages where the word Esso meant something exceptionally obscene, and so a change of name was decreed. Whatever the truth of that, its early history is pretty obscene. John D. Rockefeller built it up by coming to secret agreements with railway companies (in which he also had a high interest) to price the operations of his rivals out of the market, and by obtaining a monopoly of refining work. Other companies, unable to compete with his own massive discount railway rates were forced to come in with him, and by varying the prices in different States, once he had established a big enough network, he was able to undercut all his competitors. By 1884, Standard controlled 90% of American production. By the 1890s, he was making one million dollars a week out of exports. There was of course considerable agitation about the monopoly power of Standard Oil, but it wasn't until the publication of a series of letters which showed that the Rockefellers were blatantly evading the anti-trust laws, and trying to buy up half Congress into the bargain, that Standard Oil was finally broken up in 1909. Since then, almost every state has its Standard Oil company, while the original Standard Oil company became Standard Oil of New Jersey, and traded as Esso. Two of the fledgelings have since also achieved a place among the world's seven major companies – Standard Oil of New York, which now has many subsidiaries round the world and trades as Mobiloil; and Standard Oil of California, which began its operations abroad only alongside other big companies, but now goes it alone, and uses the Chevron name. The other two major American companies are Texaco, which obviously began in Texas, but has a large slice of the Latin American operations; and Gulf Oil which is a major participant in the Middle East. Beyond questions of who extracts what where, these companies also of course have shares in other companies operating in many parts of the globe, and major interests, or outright subsidiary companies in the refining and marketing processes.

The two non-American companies both have their headquarters in London. The first is Royal Dutch Shell, the second British Petroleum. Shell started life as a general trading company in the Far East – a sort of smaller rival to Jardine Matheson and other romantic commercial names. But when it began to transport the Baku oil from Russia to Europe, it was increasingly tempted to move into the field of oil, and after the original Rothschild contract for the Baku fields ran out, the company began to explore the possibilities of an oil trade in the Far East.

It did so largely under the energetic drive of the other great pirate of the oil business, a European counterpart to

John D. Rockefeller. Indeed, Sir Henry Deterding, as he became, was in many ways a more remarkable man than Rockefeller, a man of organising genius where the other was a business manipulator; a megalomaniac where the other was mean; in later life a Nazi while Rockefeller remained a Baptist. Deterding was probably the first man to realise the world-wide implications of oil – not unnaturally, since his own country, Holland, did not possess oil, but his company was busy prospecting for it in the Dutch East Indies. This also meant that whereas Rockefeller was pre-occupied with the destruction of his rivals in a purely American context, Deterding was trying to maximise the profitable operations of a company which had no real rivals anyway. And the maximisation of profits meant that Royal Dutch engaged in every side of the oil businesss: prospecting, extracting, transport and marketing. It was Deterding who established the subsequent character of the oil companies as gigantic organisations concerned with the entire range of the oil industry, and in a class of their own in world business.

Deterding's dream was to line up all the European companies in an alliance against Standard Oil. His capacity for diplomacy being nearly as great as his capacity for organisation, he succeeded in doing so – but the date of his success was July 1914. The victory was therefore rather short-lived. But in the meantime, he had brought about the merger between Royal Dutch and Shell in 1904, and established the giant of today.

The terms of the merger reflected the strength of the two companies at the time (Shell had been badly hit by its war with Standard Oil, whereas the Dutch company had profited greatly from winning the competition with the Germans for access to the newly developing oil fields in Romania) and the two companies united with a 60:40 ratio in favour of the Dutch. Today the Managing Directorship is still weighted on the Dutch side. Shell is also the second biggest and most complex of all the oil companies, after Standard Oil of New Jersey.

And then comes BP. As its name implies, British Petroleum is not a wholly private company: indeed, the state has a 49% holding in it, which dates from before the First World War. As such, one might think, it is unique among the major oil companies – an organisation which is not merely there to get oil from the producers and sell it to the consumers at the highest rate it can. Surely it ought to reflect the interests of the British government too. But not a bit of it. The government does appoint two men to its seven-man board, with what is called a "watching brief" in the interests of the state. But so far, this is not known to have made any difference whatever. BP operates just like any other oil company, anxious to increase the profit for its stockholders. If this profit is at the expense of the consumer (that is, the British state) that is not how BP sees it. As far as the company is concerned, its job is to provide profits, and the British government can't grumble if, like every other stockholder, it continues to draw handsome revenues from the Persian Gulf. BP has also tried, with considerable success, to take over a number of medium-sized refining and distributing companies in the United States – and though it is nowhere near in size to the three big offshoots of Standard Oil, it is now in among the major American companies. The Americans, on the other hand, have been buying heavily into Shell and Shell's American interests now account for something like a third of its total profits – a fact which is reflected in an increasingly important element of American decision in Shell policy.

These companies deal across the world, in the first place, with the oil-producing states and then with their customers, the oil importing states. But they also deal with each other, and then with a number of smaller companies which have come to life in two separate ways. The first way is that there are a fair number of American companies who were able to move in on the wealth of oil, particularly when the world seemed to have enough for the foreseeable future. They are either big enough not to be smothered by the major companies, or else they are well enough protected by American, and other, anti-monopoly legislation. They are what the trade calls the "independents" as opposed to the "majors" – though it is hard to think of anyone more truly independent than Exxon. And in the second place, the major companies have to deal with commercial organisations set up by the states in which they operate: a pattern first established by some Latin American countries, and now very widespread. They exist to protect the commercial interests of the state itself, and deal in commercial terms with the major companies; they should not be confused with the government agencies or ministries which also exist, but whose job it is to decide on the conditions on which companies can exploit the oil deposits, how much they should be taxed, and so on. In contrast, the main task of the commercial state companies is to reach marketing agreements, or agreements on refining or processing, with the foreign companies: in other words to ensure that some of what a particular country wants from its oil can be provided without the government having to step in every time; and without being subject to the dictates of a company whose interests are defined by a whole variety of calculations across the world.

But the chief question has always been, and remains: how do the major oil countries conduct their relations with the oil states? Here, the first consideration that arises is of course, that of the right of access to, and use of the oil fields. And this was arranged, in the good imperialist manner, by the system of concessions. "We", says the company, "will give you some of our money, and in return you will allow us to work your oil fields." The early concession agreements – and early, here, means right down to the period of the

Second World War – made the companies practically the sovereign masters of the territory of the oil fields. The companies were allowed to explore where they wanted. They were allowed to decide whether they wished to make use of the oil deposits if they found them. And, in return, they raised the second consideration: how much was to be paid to the state. The payment was, needless to say, small, but it was accompanied by a share in the profits. So far so good – providing the state, or anyone outside the company, knew how it kept the books. But this was not generally known, or if it was known, it was not subject to the control of the state anyway. So the share of the *declared* profits, which was at first in any case a relatively small proportion, was not a share of the *real* profits. And there was very little that the state might do about it. If it had interests in oil elsewhere, the company might threaten to shut up shop, in which case the state would lose even the little money that it did make; or, even if there was not enough Company oil elsewhere, the company frequently had the backing of the United States, or Britain. Indeed, the first major attacks on the concession system, whereby a company frequently had an oil-producing country by the throat, came, not from the oil-producing states themselves, but from other companies. This arose largely because the pattern of the concession system was established by the European imperial powers, either shortly before or just after the First World War. The companies were effectively cut in on a system of imperial control. But after the war, European states themselves faced more than one economic crisis. This made them vulnerable to American pressure, and a threat of a cut-off in the supply of American oil (still the chief source of European oil imports) would have been extremely uncomfortable. As a result, the American companies were invited into the Middle Eastern fields by those European powers which until then had had a near-monopoly. This occurred first in Iraq; later, the American companies offered to allow Shell and Anglo-Iranian (the forerunner of BP) a share in Saudi Arabia – which they refused, to their cost; during the Second World War, there was a small quarrel between Churchill and Roosevelt about the future share-out in the Middle East, but it was resolved easily by mutual assurances that each would respect the other's concessions – those of the British in Iran, those of the Americans in Saudi Arabia. The point is that all these agreements, deals, quarrels and arrangements occurred without any reference to the governments of the oil states. The concessions were there for the companies and the home governments behind them, to share out.

The concession system was very difficult for the oil states to break. They did not in fact try to do so for many years. Their first objective was to revise the terms on which the concessions worked. (This was not true of Mexico, which did succeed in nationalising its oil in the 1930s, but nobody else was able to follow their example. One reason was of course the very success of the oil companies in finding oil fields, which meant for a considerable time that importing countries could switch sources of supply, and that the chief worry of the companies was not a scarcity of oil but a glut.) But the oil-producing countries attempted to modify the system in three ways. The first was to restrict the right of the companies to explore and exploit further oil-fields. They did so by granting rights either to their own commercial organisations, or else to rival companies, including the "independents", to prospect – usually on terms which were rather stricter than those granted in the original concessions – and exploit new oil. The second way of revising the system consisted in improving the rates of payment made by the company to the state. Here, in order to get round the original notion of a share in the profits, which was seen in the end to depend on a company's book-keepers, the states began to levy a simpler tax on the barrels of oil as they left the ground.

This was a much more reliable way for the oil state to ensure its cut of the company's revenues, though it did not necessarily make much difference to the company's profits – for, in the end, the company could always pass on the price to the people buying the oil. The third way was to demand a direct participation in the board of the company – not for all its operations, but for those which concerned any particular state. The best comment on this was provided by the Chairman of Anglo-Iranian, Sir William Fraser, when the Iranian government proposed that it should have two members on the board. "What! And have them look into our books?" None the less, the oil states have by and large succeeded in amending the rules of the game to bring about all these three changes, and in this way the power of the companies over their producers – though not over their cucsomers – has declined very sharply.

The next question that companies had to consider was how they arrived at the prices for the oil, once it was delivered. Here, one might think, they had merely to consider all the factors in the cost of extracting, transporting it, and transforming it, add a reasonable profit and fix the price accordingly. But not at all. Because of the American origins of most oil companies a completely different costing was arrived at. During the 1930s, there was some danger of such a glut in oil that it would lead the major companies to engage in price-cutting wars with each other, which in turn could lead to over-production, and ruin for much of the industry. The usual solution to a price war is of course a cartel – an agreement between companies to keep prices up, and, where necessary, production down. This is what happened here. The companies did in fact agree to preserve the level of output they had already, and also to share facilities – such as refineries – where this was convenient. But the specifically American aspect to this deal was the pricing mechanism that was agreed on. The price of oil anywhere in the world was calculated as if it had been transported from the Gulf of Mexico! So that even if it were oil moving from the Middle East to Europe through the Canal, the transport was still charged as if it had crossed the Atlantic. The basis of this agreement was of course that most companies were either American, or else had investments in the American oil fields. It was therefore in their interest to safeguard American production, and the easiest way to do this was to ensure that other oil did not undersell it in a valuable market, merely because it was nearer and therefore cheaper. (This is also why I suggested at the beginning of the book that transport costs for oil, low though they were, could not be considered when comparing it with other fuels. The transport costs were artificially fixed – in every sense of the word.) The fact is that even where a company did not have extensive interests in the United States, it found this principle of pricing extremely profitable, since it just meant writing in higher prices.

This "Gulf-plus" principle, as it was known, continued until after the Second World War, and was then modified by the governmental intervention, both of the West European countries and of the United States. By then, the European countries were short of the currencies necessary to pay for the oil, and the United States saw a fair proportion of the aid it was giving under the Marshall Plan going back into oil companies' profits. Between them, they therefore managed to amend the system. But it still provided a model for other company agreements wherever they could get away with it. In West Africa, for instance, the companies agreed to take it in turns to transport the oil. A tanker belonging to one company would move down the coast, discharging the oil required into the different refineries of the different companies. But the West African countries themselves were each charged the full cost of transport, as if each company had brought the oil separately in its own tankers for the purposes of one refinery alone! *This* system was broken only when the West African states began to build their own refineries – which in their economic condition was extremely expensive.

So far, then, the workings of the oil companies went like this. They obtained concessions in oil producing states on the basis of paying the state a certain amount of money plus a share in the profits; but this share in the profits was originally low, and in any case depended on how the companies kept the accounts. They then sold the oil or oil products to importing countries at the highest price level they could achieve, including that of an artificially calculated transport system. (This of course did not show in the profit accounts shown to the oil-producing countries.) But how did they arrive at the prices they *did* show the oil-producing states?

The basis was simple. The companies "posted" (announced, that is, as if they were competing, like fruit sellers in a market) the prices they charged for crude oil. Of course, they were not like sellers actually competing in a market, since while oil was short, as demand grew all over the world after the Second World War, they were sure of a sale anyway and the price at which they eventually sold the oil was to a high degree fixed by collusion between them. But these "posted" prices became the basis on which the oil-producing *states* charged the *companies* a share of the profits. In effect the "posted" price was the oil company's tax return to the oil state, and on the basis of that return the oil state charged income tax. The system worked reasonably well so long as oil remained scarce, and so long as the oil state could gradually raise the level of income tax. But two things happened to change it. First, aware of the tremendous power of the major oil companies, the oil states began to encourage others, like the "independents", or their own state companies, to search for more oil, on a different concession basis – a higher rate of tax in the case of the independent, or with a view to taking all the profits in the case of a state company. This search was so successful by the late 1950s that more and more oil fields were found throughout the world. Equally, the major oil companies themselves were interested in prospecting for new fields in areas where they would not have to deal with strong governments like that of Iran, but where the government concerned would be grateful to have the oil in any case, and was likely to grant reasonably generous terms to the company. Hence the oil strikes in Nigeria, Libya and Australia. But the result of all these operations was that the world seemed to be about to have a large surplus of oil. And the second fact that combined with this to change the situation was the American import quota on oil, imposed in 1959, which I discussed earlier. The effect of this decision was to reduce, sharply and suddenly, the expectations of a very profitable market for imported oil – and thereby to make the surplus more serious everywhere else. Oil, in other words, was likely to be cheaper. And to some extent it did become cheaper. The United States was paying more for its oil than it need have done, but at least Western Europe and Japan began to benefit from reductions in the price of oil – a fact which stimulated their economic boom and also led them to base further growth on the prospect of cheap energy for years to come. But while this was so, and while these countries were thereby encouraged to switch even more to oil than they were doing already, and so increase the size of the market for the oil companies, it also meant that the profit per barrel of oil was going down sharply. The companies had to face a decision taken by every business man. Do you increase the turnover to sell more cheaply, so that in the end you continue to increase your profits, even though your actual profit per item goes down, or do you restrict your turnover to keep your profit per item up?

Such an agonising decision was too much for the oil giants. There was too strong a prospect that if they did try to restrict the supplies of oil to the most profitable level, other suppliers would step in with cheaper oil – the Russians or the "independents". But if they engaged in a general competition for a bigger turnover, this would really mean that their profit per barrel went down so sharply that their overall level of profits would go down too – since they could not hope to capture all the extra slice of the market which the world oil surplus now meant. So what did they do? They *did* cut the prices of the oil products to keep the market expanding, but they also reduced the "posted" prices on the crude oil. In other words, they showed a smaller income on their tax returns to the oil states, so that they would be charged a lower tax on it.

This was too much for the oil states. In effect, it meant that they were being asked to take a drop in income so that the rich American market could continue to be protected. At this point they decided to co-ordinate their dealings with the oil companies and present a united front, thus trying to make sure that they were no longer continuously out-manoeuvred by the world-wide ability of the companies to switch supplies and deliveries from country to country, thereby always obtaining the most favourable terms.

For it is important to remember that up to now, the world-wide shortage of oil had not only meant that the companies could charge high prices to the importing states. It had also worked to the advantage of the oil-producing states; for they had been able to charge higher levels of tax, in the certainty that the company would pay up, and that the increase would be passed on to the consumer afterwards. Now, the surplus of oil threatened to work to their disadvantage in the same way: they could not charge more because the companies would then switch to other countries where the prices were lower, and they would do so because *they* in their turn knew that the consumers would not just go on paying high prices if cheaper oil had become available. The only way round this was for all the oil-producing states to agree to stand together against the companies.

This was the origin of the Organisation of Petroleum-Exporting Countries (OPEC) which I have mentioned earlier. It was founded in 1960, and though it took some years to produce common policies, it began the series of major changes which have now transformed the rules of the game and revolutionised the world energy situation.

At first, all that happened was that the countries forced the companies to restore the "posted" prices to their previous level. But this was done rather on an individual basis than as united action. It was, at that stage, enough for the companies to realise that the threat of solidarity would have to be taken into account. Later, though, the power of OPEC began to show in the changes which came about in the whole character of the relations between company and state. A series of agreements increased the state's share of the profits on crude oil; and increased its participation in the management of the company – so that today the decisions taken in any oil-producing country begin to reflect the policy of that country at least as much as the policy of the company; and ensured that questions like the level of output and the pace of production would be determined by the country. That is how the wheel has now come full circle in a country like Saudi Arabia, where it is the companies who want to increase the pace of production and the state which wants to keep it at its present level. And the state is winning.

The changes that have come about, and the nature of the present situation, will be discussed in the last two chapters of this book. But it is worth summarising here the character of the oil companies of today, as they have emerged from the history of this century. No oil state now has to deal with a single company. Everywhere, the companies operate in a group, and in most countries the groups of companies represent at least two nationalities, often more. The only exception is Saudi Arabia, where the group known as Aramco consists entirely of American companies. Elsewhere, the state has to deal with either a consortium of companies which are trying to operate together, and which might represent American, British and Dutch interests; or it might deal with a series of different groups, operating to some extent as partners but also to a degree as rivals. This might include American, British, Dutch, French and Italian interests. This situation arises in the first instance from the decisions taken by the major powers to cut each other in on their concessions in the period before and after the Second World War, and which became practically universal at the moment that the United States got itself into the Iranian fields with the fall of Mussadiq. It also results from the efforts of the *importing* countries to reduce the power of the major companies, either by creating state concerns like ENI in Italy or CFP in France, or else by encouraging the "independents" as in the United States. But the point is that in dealing with the oil-producing states, the companies have to co-ordinate not only their own commercial interests but also the national interests of the countries behind them. And since the "energy crisis" of the past few years has split the importing countries very sharply, this is proving extremely difficult. The oil state, on the other hand, will belong to OPEC and possibly to OAPEC too; and though the politics of oil, which I have discussed in the last chapter, are themselves a matter of contention among the OPEC countries, the overall framework of these politics – the history of a pretty blatant form of commercial imperialism – gives them a strong enough incentive to agree on common policies.

But while this means that the power of the companies has been drastically reduced as far as the oil states are concerned, it does not mean that their profits have been reduced. Quite the contrary. In 1973, for instance, Exxon made the biggest recorded profit in history: 2,440 thousand million dollars. How can they do it?

The oil companies have become *intermediaries* between the oil-producing and the oil-importing states, and in their function as intermediaries they take a high proportion of the surplus value of the oil, the surplus value being the difference between the actual cost of producing the oil, transporting, refining and distributing it, and its cost to the consumer. Of course, the importing states also have a very large share of this surplus in the form of taxes; but so long as oil and its products are bought, the consumer pays for both.

This means in fact that the oil companies and the oil states now have a very high degree of *common interest.* So long as the oil is bought, both have the same interest in a rise in prices, because for the state it means increased revenues, and for the company it means that the profit per barrel is now very high indeed. While the profit per barrel remains high, it is worthwhile for the company to work with the oil state, and certainly more so than if it were to invest in new forms of production elsewhere. And since at present, the world has got itself into a state of tremendous dependence on oil, the market is assured. Only gradually, as oil is developed in other areas, or as other forms of energy are used, will the market begin to be more uncertain. This does give the companies (and the oil states) a problem of time scales. (How long can they keep it up? At what point should they begin to look to other forms of profit and interest?) But for the present, both have the same interest in high prices, and are likely to have it for a good few years yet.

At the second level, too, that of refinement into oil

products, the oil companies are managing to keep their profits at a very high level. They adjust the price of the finished products according to the price of the rival products that can be made from substances other than oil. But, again partly because of the widespread dependence on oil that came about in most economies when oil was cheap, rival products are still comparatively expensive. A country would have to put up a great deal of money to develop them, and, as I suggested at the beginning of the book, the very fact that oil prices have risen so sharply means that that kind of money is simply not available. In the end, then, a company is able to realise the value of its oil more quickly than the costs (to the company itself) have increased. The surplus value is greater than it used to be, and the company has a big share.

These considerations together mean that the profits of oil companies have increased enormously during the very period of the so-called crisis of the past few years. According to *Time* magazine, Exxon's profits in 1973 were 59% greater than they were in 1972; Mobil's, 47%; and Texaco's, 45%.

What does this suggest? Obviously, that the companies are no longer making their huge profits at the expense of the oil-producing states but at the expense of the oil-importing states. This of course is largely true. One might be tempted therefore to argue that all that stands between the oil producers, the oil consumers, and a reasonable policy and pricing system for the oil is the companies themselves. Get rid of the companies, and oil should be cheaper and more plentiful, and both producers and consumers would be happy. But it is not quite so simple.

In the first place, the companies do perform a valuable function as a political screen. That is to say that their very size, and the world-wide scope of their operations makes it possible for importing countries to be reasonably sure of their oil supplies even if they have serious political disagreements with the producers. And serious political disagreements are bound to arise – notably over Israel. A country like Japan which has tried to deal with the oil-producing state directly has lived to regret it for political reasons.

Second, the companies cannot eat gold. They have to put their profits somewhere, and because they face a problem of time scales, a proportion of these profits are being re-invested in other forms of energy for the future. One might argue, and perhaps argue justly, that this investment should be under the control of the state concerned. But that is a question of justice, not economics. The money would still have to be spent – and at present the companies are spending a good proportion of it.

So unless one believes in principle that profit-making companies should not exist, the arguments against the companies are not so strong as they might appear at first sight. The real argument, perhaps, is about the scale of their profits and the scale of the taxes they pay. It is quite clear that their profits are very much too big, and the taxes very much too low – especially in the United States. A heavier taxation system, which ensured a re-direction or company profits into the well-being of the state is more than necessary. Equally, it is important to keep a careful eye on the activities of the companies elsewhere in the world. The people who are paying the real price both for the policies of the oil states and for the profits of the oil companies are the poor. *That* is where agreement is most necessary, and international agreement too. But so far it seems not at all likely.

PART THREE

WHAT HAPPENED
AND
WHAT HAPPENS NEXT

5

WHAT HAPPENED

In any discussion of how an industry works in the world today, and how its products reach the people who use them, three questions are bound to arise. They are the question of economics, the question of money, and the question of politics. The first two might look as if they are practically the same, but in fact they are quite different. Economic questions are about how you produce something, and how other people value it. If it is plentiful and easy to produce, it will not be valued very highly: that is, it will be cheap – unless other things get in the way. If it is scarce and difficult to produce, it will, if it is needed at all, be valued more highly – it will be expensive. In pure economic terms, oil is plentiful and easy to produce, so it ought to be cheap. The reason that it is not depends on the other two questions, those of money and those of politics.

Questions of money are essentially concerned with matters of *translation.* How does one translate the value that one society places upon a particular product into the value that another will place on it? One might think at first that this was pretty simple. If something is basically necessary, then surely the money for it will be found, whichever society one is talking about. In which case, things ought to even out in the end, rather as in the second law of thermodynamics, which suggests that every change of temperature in one body is balanced by a corresponding change of temperature in another, so that in the end the whole universe ought to end up at the same temperature: that is, nothing will move again, nothing will happen, and the whole universe will die. But the laws of money are not like the laws of heat. Even something as basic as food can be ignored if a government prefers to spend its money on other things – or if it hasn't got the money anyway. For many years, for instance, the Soviet government preferred to starve large numbers (many millions) of the Soviet people and spend the money on building up heavy industries, either from resources inside the country, or with technical help from abroad. At the end of the period, it emerged with fewer people but a bigger power base, which is what it wanted to do. Today, the Chinese government has decided the other way round. In this sense, there is no such thing as a law of money, only a set of basic decisions. The "laws" of money only work to the extent that enough governments agree on the basic decisions – which was, for instance, roughly the case between the countries of the Western world for a generation after the Second World War. But beyond that, the cost of any particular product in a particular country will depend on how far that country is prepared to spend money on that product, and on how much money there is to spend in the first place. A developing country like India, or a country in financial difficulties like Israel, might decide that it cannot allow its citizens to spend too much money on goods which are not strictly necessary if it means importing them from abroad. The result is that a relatively cheap product like a car costs far more than its "real" price in India and Israel. Similar considerations apply to East-West trade in Europe – whose volume is largely determined either by the amount of hard currency (which is basically Western currency) that the Eastern countries have available, or by the money policies of countries which have a deficit on their balance of payments, like Britain and the United States. Both these countries have tried to restrict the import of certain products – e.g. Japanese television sets – even when these products were better and cheaper than anything they could make themselves. So, in such cases, the operation of money considerations changes the basic economic pattern.

The third question is that of politics. Again, politics represents a set of *decisions*, not a set of laws. The decisions are as likely to be based on emotion as on any set of hard-headed interests. Hard interests, for instance, might suggest that a country will get richer quicker if it engages in extensive trade with another country. But, even in peacetime, emotion can prevent this, on the grounds that it would be "trading with the enemy". The history of the Cold War is full of such examples. Indeed, one might suggest that politics is fundamentally emotional and irrational. The old definition that "politics is the art of the possible" should perhaps be changed. It is the art of making the irrational sound plausible. That is true of the history of Northern Ireland, of Britain's relations with the EEC, of the Middle East conflict, of the relations between the United States and China. Unfortunately, when the irrational sounds plausible enough, anything becomes possible, including a world war – which is why the old definition is not only inadequate but also dangerous. And in its impact on economics and money, politics can be of fundamental importance.

These three considerations, economics, money, politics, lie behind the history of what has happened recently in the relations between the oil-producing states and the oil-consuming states. In terms of economics alone, oil, as I have suggested, ought to be cheap. There is so much in the world that even if all the countries in the world go on increasing their use of oil and their dependence on oil at the present rates, there would still be a surplus ten years from now – and that is only from the proven oil reserves. Nobody believes that these proven reserves really represent all the oil there is left. Indeed, they might only be a fraction. But oil has become vastly more expensive because of the other considerations. And here, there is a paradox.

Most countries, most of the time, have decided that if the import of a product strains their money policies too tightly, they will cut down or go without. They will not allow all their money to disappear into the hands of foreigners; they will be determined to have money available for other things. The present situation is the reverse of this. The advanced industrialised countries of the world have given no indication that they are going to cut down on their use of oil. Instead, they are pledging themselves to increasing strain on their balance of payments – perhaps in some cases pledging themselves to bankruptcy – in order to keep up their imports of oil. In other words, it is *the money policies of the oil states,* not the money policies of the importing states which now determine the price of oil – with the companies helping it along, of course. This is the opposite of the usual situation. Normally, it is not the Germans who determine the price of a Volkswagen in India but the Indian government. Why has this reversal taken place? The answer lies in the politics. In discussing the politics, I shall try to say what happened as plainly as possible. But it is worth pointing out that there is no agreement on this subject, even among the experts. So I shall also say where the controversies are, and in doing so, I shall return briefly to some questions I have already sketched in the chapters on the oil states and the oil companies.

But the first basic characteristic of the situation which has developed now is that the politics of oil have become internationalised. This might sound like a cliché, but it is not. For the pattern of international politics which has emerged in the last four years or so has succeeded a previous pattern in which the different oil-producing countries were trying to change their relations with the oil companies in order to promote *internal* changes in their respective systems. When this was so, they came up against the fact that other countries were not necessarily sympathetic to these changes; and in consequence it was possible for the oil companies to outmanoeuvre them through arrangements elsewhere. The classic example of this is the fate of the semi-revolutionary Tudeh party in Iran which chose a confrontation with the Anglo-Iranian oil company as part of an attempt to force the Shah to change the system of government. In the end, it failed on both counts because Iran was faced with bankruptcy as the oil companies opened up other fields in the Persian Gulf with the blessing of governments which did not approve of the Tudeh. Today, it doesn't matter whether the oil companies are dealing with a conservative monarchy in Saudi Arabia or a revolutionary government in Libya. *All* the oil-producing states have a basic strategy in common, even though they might differ on questions of prices or production at a particular moment.

This means in turn that there is a second characteristic of the new situation. The strategy of the oil states is not directed so much at the oil companies but at the oil-importing *countries.* This might lead one to suppose that there could be, or should be, a common response on the part of the oil-importing countries to the attempts by the oil states to make them pay more than they can afford. Indeed, this was the assumption behind much American policy during the six months between the war in the Middle East in October 1973 and the conference of the oil-importing countries called by President Nixon in the spring of 1974. But in fact, it didn't work like that. The basic situation of the different countries varied too much. For while, for instance, the United States needs only to import six per cent of its energy requirements from the Middle East, the Western European countries rely on the Middle Eastern fields for up to 80% of their oil requirements. The United States, therefore, could afford a relatively tough line, and tried at first to persuade its allies to adopt one. But, provided that they were not going to cut down on their own oil imports, neither the Europeans nor the Japanese could afford any such thing. So, the internationalisation of oil politics has meant that, as the oil producing states have achieved greater solidarity in their dealings with the rest of the world, the major oil consuming states have become more divided. It is largely for this reason that the immense price increases of recent years have been so successfully imposed.

The third characteristic follows to a degree from the second. It is that the solidarity of the oil states depends very largely on the fact that so many of them are in the Middle East. I have already suggested that this is where the greater part of the oil reserves are, and I have shown to what extent Western Europe, in particular, has come to rely on the Middle Eastern fields. One figure alone is enough to make this even clearer. The proven reserves in the Persian Gulf area amount to 367 thousand million barrels. Even in 1985, and even if the world-wide demand for oil goes on growing at its present rate, the world's oil imports from that area are not likely to be more than 15½ thousand million barrels a year. In other words, if the Middle Eastern states are sitting on a fantastic oil surplus, they are in a position to make the running on prices and production. Other countries – Indonesia, Venezuela, Nigeria – will follow suit. And it is on the conflicts and tensions of the Middle East that the politics of oil has come to depend. For the Middle Eastern countries all have some reasons – rational or irrational – to wish to revise their relations with the Western world, and to this end, they have in oil an admirable instrument. And just as their solidarity, compared with the divisions among their customers, *enables* them to impose new prices, so does their history give them the motive to do so.

These three characteristics, taken together, give some sense of the underlying structure of the changes that have been going on in the past few years. But it is at this point in any discussion of the subject that one begins to hear the words "oil weapon". And this is where one of the major controversies comes in. The "oil weapon", as such, suggests a war – and there *is* a war. It is the permanent war between Israel and the Arab states, which sometimes erupts into fierce fighting. Since Israel always wins the fighting, or is in a position to do so by the time it is stopped from outside, it has been natural for the Arab states to try to use the other weapons at their disposal to make up for their defeat. And in the view of some analysts of the situation, the attempt to use the oil weapon against Israel, by threatening the Western supplies of oil in order to make the West put pressure on Israel to retreat after its victories, is the *prime* factor in the changes that have occurred. A very distinguished analyst of the oil industry, Professor Edith Penrose of the University of London, argues for example that this is the case. I personally believe that other motives – the desire in some countries to gain more money for development, the revolt in all countries against a pattern of imperialist domination, the sense of a new Arab potential for power in the world – have been more important, and that the one point on which the Middle Eastern states have shown very little solidarity is on how far they should go in trying to use oil as a weapon in their conflict with Israel. But more important than the controversy about the degree of importance of different motives is the fact that a term like "oil weapon" can lead one to overlook the real nature of the changes which have occurred.

For the changes have been in the context of the relations between the major international companies and the oil states; and of the seven major companies, five are basically American, and American interests have a very substantial holding in a sixth – Royal Dutch Shell. Now the United States, while it is the only country in a position to put much pressure on Israel, is of all countries the one which is least vulnerable to any Arab threat to its oil supplies. The Europeans and Japanese *are* vulnerable, but they are in no position to affect Israeli policy. And, as the recent history of relations between the United States and Europe has shown, the United States is in no mood to listen to European representations when it feels that European governments are betraying the solidarity of the West.

Indeed, one might go further. Before the use of the "oil weapon", in 1973 and 1974, the American dollar had become a pretty weak currency in comparison with its European and Japanese rivals. The result of the immense price rises which the OPEC countries brought about, though, had the effect of weakening the European currencies to the point where the dollar emerged once more as internationally the strongest. In this sense, the Americans were every bit as much the beneficiaries of the "oil weapon" as the Arabs themselves. Instead of having to beg or bluster their way to Brussels, in an attempt to negotiate some revision of the international money system with a group of strong European countries, they recovered something of their former position as the managers of the world economy. So why should *they* respond to the oil weapon by putting precipitate pressure on Israel?

In other words, if the main Arab preoccupation in bringing about changes in the oil situation were to pressurise Israel, their policies would be a total failure. In a wider view, however, it has been a considerable success – partly, of course, because of the wider long-term interests of the United States. More than that, the view that changes in oil policies are primarily an "oil weapon" leads to an even greater misunderstanding. That is of the fact that the Arab oil states and the oil importing states, especially those of

Western Europe, have considerable interests *in common.* I have already suggested that Arab, and other, oil-producing countries, need to regulate the pace of oil production, so that they do not turn a real asset – oil in the ground – into a wasting asset – money in the bank – too fast. But this in fact suggests a much deeper common interest with the consumers of their oil. For the pricing as well as the pace of oil production is almost as much a matter of concern to the oil states as it is to their customers. If they raised prices so high that they drove the Western countries (as they very well could) into bankruptcy, it would mean that all the revenues they are enjoying from the sale of oil would suddenly become worthless, and they themselves would also suffer heavily. For this reason, they need to be careful about the level of the prices they charge – as much as they can get, certainly, and as much as their customers can pay without facing ruin in the long term, but no more than that. Both the sellers and the buyers of oil might be wrong about how much the buyers can afford to pay; but at present there is a more or less agreed level, and it might even begin moving down, even before other oil reserves in the world are opened up. If it does move down, this will be because the oil producers, and especially Iran and Saudi Arabia, judge that it is in *their* interests that it should.

In fact, there is a further, though seldom expressed interest which is common to both the oil states and their customers. It lies in the possibility that if the Western countries became desperate, they could try to take over the oil states by force. Such an eventuality has been mentioned once or twice by the governments of the oil states, and they have threatened to blow up the oil wells if it happened. (Though one can of course put fires out.) But more generally any such attempt could either mean a world war, which would be in nobody's interest, or it could mean a revival of the history of Western imperialism – which would be in nobody's moral interest either, but would certainly be in the material interest of the advanced and desperate Western states. And for a successful imperialist *coup*, as opposed to a world war, all one would need would be the agreement of the Soviet Union – perhaps with a Soviet share in the proceeds. At present this is enough to rule out such a prospect. But it is not altogether fantastic to suggest that the Soviet government is quite capable of changing its mind about whom it would collaborate with. Stranger things have happened in the short history of that country; and if Soviet interests were threatened – in the procurement of technical help, or long-range credits from the Western world – because the Western countries were either going bankrupt or running out of oil, such a change could come about. The possibility is still entirely theoretical, but it has certainly been discussed by some of the Arab governments as a long-term danger and obviously as one which they would wish to avoid.

So, for political as well as economic reasons, there is a higher degree of common interest between the oil states and their Western customers than might appear at first sight. What one is considering, therefore, is not really an "oil weapon", except at very specific moments of Arab-Israeli conflict, nor an all out struggle between the Arab states and the Western world, but a finely balanced set of tensions, in which short-term competition for money and power must be weighed against an underlying and long-term need for co-operation. If this is so, it means that what has happened in the relations between oil states, companies, and oil-importing states is an attempt to change the terms on which oil is sold, so that the exporters enjoy more money and more power than they have had in the past.

By what means, then, has this transformation been achieved? The first, and most essential point is that OPEC began to work. For some years after its establishment in

1960 – which took place, one might recall, as a reaction to the American creation of an oil surplus in the rest of the world by restricting American imports of oil – OPEC did what international bureaucracies generally do: it elected a Secretary-General, established headquarters in expensive offices, and talked to itself. But there was a good reason for this inactivity. The purpose of OPEC was to maintain "price stability", in its favourite phrase, and for some time this meant an attempt to prevent prices going down as a surplus continued to threaten. Now, as one might immediately guess, there were other people who shared exactly the same interest in price stability as OPEC, namely the oil companies. And so, though on the face of it, OPEC and the oil companies were mutually hostile bodies, the one representing the interests of all the oil states, and the others representing the interests of their own shareholders, they were in fact allies. Both faced the same threat of a price cut, and neither wanted it. In general, the oil companies actually welcomed OPEC for these reasons, though of course it would have been impolite to say so out loud. But in private, so long as the oil states continued to be content with the "posted price" system of taxation, relations between the OPEC countries and the oil companies were excellent.

It was when the "posted price" system was threatened by the developments I have described in the last chapter, and when the OPEC countries began to see that "price stability" could mean something entirely new, that OPEC really became an organisation with teeth. The something entirely new was of course the realisation of the fact that prices could actually be increased, oil surplus or no oil surplus. After all, the "laws" of supply and demand again only worked so long as governments were agreed on basic decisions: the basic decision in this case being that if the oil was there to sell, it should be sold. Once this decision was broken in favour of a different argument – that oil would only be sold at the pace the oil-producing country wanted – then the laws stopped working. Prices could increase. This is what began to happen in 1970. At first, the decision by the OPEC countries, and in particular by the Arab states to bring about a price increase took the form simply of demanding higher taxes from the company. Whether the companies then passed on the higher taxes in raised prices to their customers was no concern of the state itself. The decision was also powerfully helped, at least psychologically, by the fact that at this time the United States was beginning to worry about an "energy crisis."

This term, like "oil weapon" again conceals a controversy, but it is one that I shall go into in the next chapter. For the moment, its importance is that the United States had begun to view with some alarm the prospect that it, and perhaps the world more generally, could be running out of energy by the late 1980s. Allied to this was the prospect that whatever the restrictions of the past, the United States would have to import very much bigger quantities of oil from the Middle East if it were not to experience an acute shortage. The President had established a special body – the Presidential Task Force – to report on the subject, and the Task Force very sensibly refused to panic. It reported that the country could easily get by on a level of 5% of imports from the Middle East, and that prices could be expected to remain reasonably low. But, on the one hand, the prospect of an energy crisis some time in the future gave added impetus to the reflection that oil prices *could* go up, and that "price stability" could be concerned with something more than simply maintaining them; while on the other hand, the notion that prices would remain low was distinctly unappealing to those countries which wanted to increase their share in the revenues to be derived from their oil. The result was that very soon, the idea of an "energy crisis" became transformed into the much more specific one of an "oil crisis."

The reason was simple. If OPEC's teeth began to bite, the oil countries could decide on their own how much oil would be produced in any year. They could drive prices up by producing a real shortage; if they were not to do so, higher prices had better be paid anyway. Hence, the *politics* of oil could transform the *money* situation, whatever economics were involved. And it is worth pointing out here that many of OPEC's activities are quite explicitly motivated by political considerations. OPEC includes most of the world's oil-exporting countries, ranging from Saudi Arabia, through countries like Venezuela, Nigeria and Indonesia, down to the tiny sheikhdoms of the Gulf. But it does not include all the exporting states: Trinidad, for instance, was blackballed by the OPEC club because other members objected to its political position. Similarly, it is very unlikely that even if some European countries do become net exporters when the North Sea oil starts to flow, they will be allowed to join OPEC. Such political considerations had their effect from the beginning: in order to avoid artificial shortages, the oil companies would probably agree to pay higher prices; and this is what began to happen in 1970.

In January of that year, Libya demanded very heavily increased taxes, and by April Exxon had agreed to pay them. This was one of those instances where the fact that an oil company is vulnerable at the fringe of its operations, in the way I suggested earlier, can make a great difference to the more general situation. The complexity of the company's operations was such that at this time it was temporarily short of tankers for the collection or delivery of oil elsewhere in the world. It needed Libyan oil badly – so much so that it was cheaper to pay the higher taxes than watch its profits go down through non-delivery of the oil. Hence the surprisingly ready agreement. But with the example of the world's biggest company behind them, the others were less able to resist pressure from the Libyan government. A few months later, in September, all the American companies agreed to pay up. This was only a start, because it began to lead now to a contest among the oil states to see how far they could make the companies pay higher taxes. Iran, in February 1971, managed to secure a somewhat better agreement than Libya had done, and for the first time "price stability" was officially translated into an agreement to pay higher prices. The agreement signed with Iran was in fact supposed to stabilise prices for five years, merely making allowance for the rate of inflation – that is, prices *would* go up in that period, but only at a rate in keeping with the general inflation rate. Real prices would not increase. That was supposed to help to stabilise prices more generally, and it lasted exactly two months. For in April, Libya again pressed for improved terms, and ended up with something better than Iran had got; and not only that, but inflation was now seen to be running at a much higher rate than had been expected anyway. Special agreements were later signed at Geneva with the oil-producing states to deal with the problems occasioned by this new development. But before that, the oil countries were beginning to change the nature of their demands.

They were by now no longer merely interested in getting more money out of the companies: they wanted to take an active part in the decisions on production itself. The simplest way to do this was to demand an equal share in the operation of the companies in the different countries. (This did not mean, of course, that the OPEC countries as a whole wanted half shares in any of the companies as a whole. It meant that in the operations of any one company or group of companies in any one country, that country should have a half share in those operations.) This principle – of equity participation, to give it its official name – was demanded by the countries around the Persian Gulf, under the leadership of Saudi Arabia, in July 1971.

And by March of the following year, the oil companies had accepted it in principle. (In fact, the Oil Minister of Saudi Arabia, Sheikh Yamani, later confessed that he had been surprised by the ease with which the companies yielded. But profits are profits.) And in the autumn of 1972, a General Agreement on Participation was signed, which provided for the individual arrangements in the different countries. Government participation was to rise from 25% to 81% over the rest of the decade. In the end, only three countries did accept it: first Saudi Arabia, and the tiny sheikhdom of Abu Dhabi, followed afterwards by Qatar. Iran and Iraq, for reasons which they were soon to show, were not interested, and in Kuwait, the Assembly, with a increasingly common show of independence, refused to ratify an agreement which the government had already reached. (Kuwait was shortly to show even greater independence.) None the less, even an equity participation agreement with Saudi Arabia alone, was enough to change the situation fundamentally.

Now, it might be argued that the companies made a mistake here. Their interests after all do not cover only the operation of taking oil out of the ground – which is what the half-shares principle was concerned with. They extend to transport, refinery, distribution, and all the rest. Obviously, if the supply of oil is threatened with interruption, and the profits are therefore liable to drop, it is tempting for them to give the oil state what it wants. But if, on the other hand, the oil companies had allowed their interests in the oil states to be taken over, they would still have maintained very profitable organisations throughout the rest of their line of business. (The "downstream" operations, as the oil industry loves to call them.) If they had continued with the downstream operations, but abandoned the business of extracting the oil, they would immediately have found themselves in the position of being the principle *customers* of the oil states. As customers, they would have been able to say what kind of price they found acceptable, and what kind of level and pace of production they could deal with. In short their position could have been that of a powerful negotiator. Instead of which, they found that, through accepting the equity principle, they were involved from the beginning in taking decisions about the level and pace of production, the marketing prices, and so on. In consequence, *by participating,* they found that their bargaining powers were much reduced. And they were reduced in the very country – Saudi Arabia – which contains most of the world's cheapest oil reserves, and which is able to take decisions affecting all the others. So, the equity participation principle which they accepted turned them from customers into accomplices in the oil policies of Saudi Arabia, and deprived them of their bargaining position.

This argument in fact concerns a further controversy among the professional analysts in the field of oil. Professor Adelman, whom I have already mentioned, has eloquently argued that the trouble with the linked questions of oil prices and the level of oil production is that the oil companies simply get in the way. In his view, the companies should be persuaded to get out of the business of the production of oil altogether, and concentrate purely on their downstream operations. In that case, they would have the maximum incentive to make the profits in transporting, refining and distributing oil and oil products. This would mean that they could no longer afford to buy very expensive oil, and they would therefore have to withstand the attempts by the oil producing states to raise the prices still further. Whereas, if they do continue to participate in the production of oil, they also have an interest in high oil prices, since these help to maximise their overall profit. The argument goes, therefore, that in spite of their conflicts with the oil-producing states, since the OPEC countries began to acquire some teeth, the companies are now basically hand in glove with the oil-producers, and now make their profits at the expense of their customers in the West.

It is a persuasive argument in some respects. But it does seem to overlook the political motives that the oil states might have in bargaining with the countries which need to import their oil; it equally overlooks the fact that some of the governments that control so many of the world's oil reserves – like Saudi Arabia and Abu Dhabi – don't actually need any more money at all. That is, they could afford to stop production altogether, were it not for their own demand for greater power in the world, and their underlying sense of a fundamental common interest with the West. (Except, of course, in the case of Israel. But here, my whole suggestion is that Israel is an *interruption* in the politics of oil, rather than the real motive behind the way the politics work.) So, all in all, what one might say at this stage of events is that the Arab states in particular, began to realise their power to modify and improve the terms on which they sold their oil to the companies and the advanced industrial world. That, having done so, they tried to change the terms so that they could improve their own financial return on the sale, and also create a new structure of power in the oil industry, one in which they were enabled to take decisions themselves, instead of being subjected to the decisions of foreign boards, whose books they could not even look at. And in doing *that,* they could begin to formulate a long-term approach to the problems of raising and selling oil. This long-term approach would both give them some chance to redress a balance of power which had been so humiliating for them in the past, and also, in the case of some states at least, give them the money and the terms by which they could develop their own social and technical revolutions.

But what this also means is that there was a fundamental split between oil states which concentrated on one set of aims, and those which concentrated on another. Earlier in this chapter I argued that all the oil producing countries had a strategy in common. In the sense that they could all agree on the need to raise prices, on the implicit threat (and sometimes the explicit threat) of slowing down production or cutting off supplies, this is true. In the sense that they all want the same thing as a result of the strategy, it is not true. Some want to raise production along with prices; some want to slow down production and raise prices. Saudi Arabia sometimes seems even to toy with the idea of keeping production at about its present level, but lowering prices a bit in the meantime. I have already discussed some of the reasons for these differences. Those countries which have large and poverty-stricken populations need money faster than those countries which don't even know how to spend the money they are getting already on their tiny populations. (The Sheikh of Abu Dhabi for instance, is in the occasional habit of sending lorry-loads of free chocolate bars round all the schools in his sheikhdom as one way of spending the money and showing his concern.) Now, the countries with a more urgent need for money have shown less interest in slowing down supplies of oil but more interest in raising prices. Those countries with a less urgent need have shown greater interest in slowing down supplies but also a smaller interest in raising prices. Into the first category fit Iran and Iraq; into the second fits principally Saudi Arabia.

And Iran and Iraq were the two states which refused to touch the participation agreement of 1972. The reasons were made apparent later, when in June of that year Iraq nationalised the operations of the oil companies at one of the country's two major oil fields – that of Kirkuk. Needless to say, this action led to a prolonged dispute, which was finally settled in March 1973; and in July of 1973, Iran nationalised the oil industry in the entire country. The companies were now permitted to raise the oil, on a twenty-year contract which gave them permission to get at the oil, raise it and sell it; but after 1993, even if the agreement holds in our present unpredictable world, the country will be wide open to whichever organisation the government prefers. This looks therefore as if the companies have lost heavily in their relations with Iran and Iraq. In a sense, they have. But in another sense, they are also reaping the benefits of those countries' refusal to accept the participation agreements. For there, there is no interest in cutting down the supplies of oil. On the contrary, both states were notable, in the aftermath of the Middle East war in October 1973, for their refusal to suspend production or cut down supplies of oil. They *were* of course interested in raising prices, especially Iran, but this was of little account to the companies. On the contrary, the companies have continued to make more money anyway.

This suggests why it might have been a mistake for the oil companies to accept the participation agreements precisely in those countries which had no interest in keeping up a steady pace in the production of oil. But it strains the chronology of what happened a little. To get back to the period before the October war: Western Europe was being gravely warned by the United States in the spring and summer of 1973 that the "oil crisis" was now upon it. The reasons were that the United States, if it wished to keep oil prices lower than they might otherwise have risen, was certainly not disposed to import even greater quantities of Middle Eastern oil. This meant that its own oil reserve was in danger of falling low, and that it could not therefore afford to keep any strategic stockpile ready for Western Europe, if the Arab states should ever again cut off supplies – as they had done in the Middle Eastern wars of 1956 and 1967. In which case, Western Europe would itself be running dangerously low. This was the nature of the "oil crisis" which took Western Europe so much by surprise during the period after the October war, and which has prompted so much writing about oil. But it should be stressed repeatedly, that this was not a crisis in the sense that Western Europe was ever in danger of running out of oil. It was a crisis in the sense that the political and financial arrangements on which the supply of oil depended would have to be revised.

The sense of impending crisis was made worse by the fact that the United States itself seemed to be running out of oil in the summer of 1973. This was the famous period when gasoline stations were shut down across the continent, and there was a certain degree of panic in the air. But all that actually happened was that the oil companies were encouraged by the situation to bid higher prices even for "participation" oil in the Middle East, and that there was a degree of chaos in the oil-markets, shares rising steeply and falling abruptly, and so on. But at no time was there a real oil shortage. Even during the height of the American alarm, it was clear that temporary shortages were due to domestic confusion rather than to the evil designs of the Middle Eastern states. *But* it did mean that the Middle Eastern states were encouraged to press their advantage while it lasted (not all of them, in fact only one oil-producing state knew that a war was coming in October – the rest were just making hay) and they therefore drove the hardest bargain they could. It suddenly became clear to them that what they had thought of as famous victories – the greatly increased prices and the participation agreements – were only a beginning. It was generally acknowledged by September 1973 that the Teheran agreement which was supposed to have provided for a five-year period of stable prices was already out of date and the best that the oil companies could hope for would be a general agreement to raise prices once more, rather than a competition among the individual states to see how far they could go.

But such hopes were shattered a month later. In October, the combined forces of Egypt and Syria, soon to be helped by those of other Arab states from Morocco to Iraq, attacked Israel. The bitter and bloody war which followed had two important effects on the politics of oil. First, the demands it made on Arab solidarity led the Arab oil-producing countries to agree on a common policy towards the principal supporter of Israel – the United States. The policy was to suspend supplies of oil to that country, and to any others which appeared to favour Israel. This embargo appeared for a time to be a very significant factor in oil supplies, for if it had worked or been continued, it would have meant a genuine shortage of oil in some areas of the advanced industrial world, a shortage whose consequences could not be foreseen. But in fact the United States is not only a close supporter of Israel; it also has excellent relations with some of Israel's enemies, notably Saudi Arabia. And the embargo was circumvented by the oil companies in the case of other countries like Holland; in the case of the United States, American diplomacy and Saudi agreement proved strong enough in the end to have the embargo lifted. There was not, in fact, a real shortage of oil at any time. (Those who remember the panic queues of motorists at filling stations in England and elsewhere, or the no-driving rule on one day a week imposed in other countries, might be surprised at this. But if they are, then, they do not remember what real shortages are like and petrol for cars anyway accounts for a very small fraction of oil imports. Even rationing did not happen.) *But* the embargo was important in the sense that, by creating expectations of a shortage and with that an economic slump, it made importing countries, and the oil companies which served them, very much readier to pay higher prices for their oil.

And the second, and real, effect of the war was to make

it possible for the Arab oil-exporting states simply to break off negotiations on raising prices, and announce a massive increase, take it or leave it. This decision was not reached by all the OPEC states, it was the Arab club of OPEC members, or OAPEC, which had been founded after the Middle East war of 1967, that took it. The price increase was a straight jump of 70% of the prices on which the oil states charged the taxes to the oil companies. Even this was not the end. During the months of confusion, bargaining and auctioneering which followed, some oil states – Libya, Iran, even Venezuela – were able to press for agreement on even higher prices, or announce that they would impose them anyway. By January of 1974, prices had reached a level that represented 400% of what they had been a year earlier. Indeed, less than a year before, one influential American article had warned that the price of oil might rise to $7 per barrel. This was thought to be unduly pessimistic at the time, but by January it had risen to $11.65 a barrel. Out of this price, the oil state gets about $7. The countries which import oil in such massive quantities foot the bill. In prices and taxes, the individual in these countries now pays so heavily for oil that it represents a real threat to his standard of living – though so far this threat has materialised only in the poor countries of the developing world (which is no longer developing at present) and to some extent in Japan. But everywhere, the expectations of a continuing rise in the standard of living have had to be scrapped; and in most of Western Europe the only real question is whether the present standards can be kept up for much longer.

There is, understandably, a degree of bitter resentment in Western Europe at the thought that the $7 which goes into the pockets of an Arab ruler for every barrel imported by the West should make him so rich, and apparently so pointlessly rich. Saudi Arabia can now afford to buy up General Motors once a year, yet only a proportion of the small population of that country seems to benefit from its enormous wealth. The heads of tiny kingdoms have taken over, or have acquired very heavy stakes in London properties and London companies, like Commercial Union Insurance. Banks in Switzerland are on the look-out for anyone willing to borrow a few million dollars, and it invariably turns out that behind this quest for custom there sits an Arab prince or an Arab minister. But, while it is true that in some cases not nearly enough is done to turn the oil revenues into benefits for the society of an oil state as a whole, it is also true that in other cases, this is just what the oil revenues *are* being used for. And in those countries which desire the money for their own internal development, there was considerable pressure for even higher oil prices.

In a sense, $11.65 a barrel represents a victory for moderation. At the end of 1973, the Shah of Iran was reported to be pressing for a rise of up to about $17 a barrel. It appeared that in his view, the prices should be fixed at about the level at which it would cost the Western world to develop alternative sources of energy. It was the Arab states, led by Saudi Arabia, which argued against this, showing at least some awareness of their underlying common interests with their customers, and a sense of the dangers that could follow from Western bankruptcy.

But while the rise in prices was less steep than it might have been, the nature of the participation agreements changed fundamentally during 1974. Here the way was led by Kuwait, whose Assembly had previously rejected the old General Agreement negotiated in 1972. Under the terms of the 1974 provisions, the government of Kuwait was to acquire a 60% controlling interest in the rights, operations and existing facilities of the Kuwait Oil Company in Kuwait. (True to the common form, the Kuwait Oil Company itself represented an arrangement between the separate oil companies: in this case, the KOC was 50% British Petroleum and 50% Gulf Oil.) Under the terms of the new agreement, the government and companies were each to determine, on the basis of their respective shares in the new structure, how much oil was to be lifted from the ground each year. The yield from their operations could be purchased by one party from the other at an agreed price: the buy-back price as it is known in the trade. Finally, the whole would be run by a Joint Management Committee, whose decisions depended on a 75% vote. Since the Kuwait government was to have 60 votes, and the companies 20 each, this effectively meant that the government could not act without the consent of the companies, provided that they agreed with each other.

The importance of the new agreement was not that it was implemented: it was not. Once more the Kuwaiti Assembly threw it out, largely on the grounds of the provision for a 75% vote; instead it demanded full nationalisation. Its true significance lay in the fact that the previous gradualist and moderate approach favoured by the Gulf governments in 1972 was now shown to be politically so out of date as to be worthless. From now on, the only question about participation agreements would be not how big a proportion would the oil states take, and how soon, but how soon would it be before they acquired a 100% participation in the different areas, and reduced the oil companies to transporters, refiners and salesmen? By the end of the year, the answer was clear: Saudi Arabia agreed with the consortium of American companies there (Aramco) that it would indeed acquire a 100% interest. As I have already suggested, this does not necessarily work to the disadvantage of the companies. If they are let off the hook of participation in lifting the oil, and restricted to downstream operations, they can even improve their bargaining position.

But, once more, what needs to be remembered is the political motivation behind the shape of the new agreements as they emerge. In the case of Saudi Arabia, it might be true, for example, that the 100% participation represents a political victory for the forces of nationalism but equally a victory for those who favour keeping prices at their present level. It does after all make it harder to increase prices greatly if one is negotiating with companies who stand to lose from a rise rather than gain from it. But other countries might not see things that way. They could regard their own control of the oil-lifting operations as merely the preliminary to another rise in prices; and this is the kind of difference one should always bear in mind when discussing the politics of oil. One indication as to what is afoot in any particular situation is the buy-back level agreed on in those countries where there is still a mixed state-and-companies participation. A very high level of buy-back price would mean either that the country concerned is not interested in marketing its oil at the present level of prices, or that it is trying to restrict the companies' own operations. It might even mean that it is also trying to replace the company in some of its downstream activities. But a relatively low level of buy-back prices would mean that the state is concerned with the pace of production and with price stability meanwhile; in other words, that it is looking after its own interests, but not waging war against the Western economies or against the oil companies as such.

The situation is obviously open to change. But it is clear so far that two of the most feared dangers that could have arisen from the changes of the past four years have not come about. The first was that the oil states might decide to cut down on supplies of oil generally. In fact, the reverse has been the case. During the period of very rapid price rises in the first half of 1973, the output of Iraqi oil, for instance, increased by more than a third, compared with the same period the previous year, and so did the output in Saudi Arabia. These levels have been maintained

– which means that they are, even today, far in excess of the rise in the world's demand for oil, which runs at about 7% a year. The other danger that was feared, especially after the October war of 1973, was that the Arab states might start discriminating in their supplies of oil on political grounds. It is true that they did start, and in some respects their behaviour was pretty outrageous by normal international standards. Not only did they declare the oil embargoes in the first instance, they also demanded that Japan, for instance, should break off diplomatic relations with Israel altogether if it was to go on receiving oil. With a rare display of firmness, Japan refused, though it did eat some nauseatingly humble pie in the language it used to do so. The oil states also demanded – and got – a promise from Britain and France, competing at the time for the title of the Arabs' best friend in Europe, not to allow any of their oil supplies to reach Holland druing the period of embargo. And even the embargo against Holland itself was an incredible act, since Holland was no different from any other European country in its attitude to Israeli policy. The Europeans all supported Resolution 242 of the Security Council, calling on Israel to withdraw from territories it had occupied in the war of 1967. But the difference was that some Dutch politicians had had the cheek to express some sympathy for Israel when it was faced with a massive surprise attack, and that was enough. So there were some grounds for thinking that the Arab states might proceed to relish a form of self-righteous but irresponsible political power. But it has not worked out like that. Embargoes have been dropped, and supplies kept up.

It was also for this reason that it seemed to me impossible to argue that the question of Israel, or the use of the "oil weapon" were the primary motives of Arab countries. Israel has been an intrusion into the more general politics of oil; the war of October 1973 certainly gave an important impetus to more general developments. But the developments have continued to bring about a general revision in both prices and participation, without Israel continuing to be a factor at all.

What I am suggesting in general therefore is that there has been a revolution in the relations between oil states, the companies and their customers over the past four years. This revolution has had two sides. The first is in the power of the oil states to determine their own oil policies – how much to produce, how to regulate the pace of production, and how much to leave in the ground. It has turned them from the virtual colonies of the oil companies which they were even at the end of the Second World War, into very powerful independent political forces. The other side of the revolution is in prices. The present price of oil might not rise much, and might even fall somewhat. But it is such a major change that the economy of the whole world is affected, and the economies of Western Europe and Japan have been drastically changed. Britain, to take an obvious example, is now borrowing at the rate of some 5% a year of the value of its whole production, simply to go on importing oil. No country could keep this up for very long: in Britain the hope is that it can begin to recoup after three or four more years by the exploitation of North Sea oil. (In fact this is only a hope. Many people have now argued that Britain is so heavily in debt that it has mortgaged all the profits it might expect to have had from the North Sea, and might even never recover.) But though Britain is in some ways the most obviously affected country (since it is straining to import oil, *and* develop the North Sea field at the same time) all the advanced industrial countries have been affected. The United States can of course survive. A very small reduction in its prodigious consumption of energy could make it practically independent at least for the time being, of its imports from the Middle East. But the other countries of the Western world are not now in a position to survive without maintaining reasonable relations with the Middle Eastern states.

Their hope is that they will be able to. This is the importance of the "Euro-Arab dialogue" that I wrote about earlier. There are two aspects to such an arrangement. The first lies in the hope, widely expressed in Europe, that those oil states with a vast surplus of Western money will re-invest it in the further development of Western Europe. The oil states would of course make a profit – though probably not of the order of those made by the Western oil companies out of *them* for so long. But the money would come back into Western Europe, and provided that the money comes in, it doesn't matter much whose pockets it comes out of. As a matter of fact, and to keep the record straight, most Arab oil revenues from Europe never leave that continent anyway. They remain in European investments. "Re-cycling" is a fashionable, but rather silly word. There has seldom been any question of bringing Arab money *back* to Europe; it is the power which the Arab nations acquire from leaving their money *in* Europe which is at the root of the problem. In this sense, the popular resentment which is sometimes seen in the press against Arab investments in Europe is exactly the reverse of what many thinking Europeans feel Europe's real interests to be. These real interests are also opposed to what much American diplomacy seemed to desire early in 1974. President Nixon's attempts to convene a conference which would create a club of oil consumers to deal as a bloc with the oil exporters (a kind of OPEC in reverse) was not only disliked by the Arabs. Some Europeans also resented it, as an obstruction to their attempts to improve relations with the Middle Eastern countries. Since then, a most reasonable compromise approach has been worked out which I will discuss later. In March 1974 the members of the European Economic Community decided in principle on a scheme of co-operation with the Arabs. It still has to be worked out in detail; but it immediately raises a second kind of interest involved in a "Euro-Arab dialogue".

This goes beyond the maintenance of a workable money system – desperately important though that is. It is probably true that unless enough Middle Eastern funds remain in Western Europe in the form of investment, the countries there will face ruin, which is clearly in the Arab interest to avoid. But the OPEC countries also have other interests. First, they have reason to believe that their present power and wealth cannot last for ever. As other sources of energy are developed, the industrialised world will come to rely less on oil of any description for the present price of oil makes it more or less worthwhile for the advanced technological countries to try to develop other sources. (Their nature will be discussed in the next chapter.) It is therefore in the interests of the oil states, not only to sell oil at the best price and at the best pace they can arrange at present, but also to become involved in the development of other energy sources. They are now in the position of investors who know that their money has to be ploughed back into new industries, to ensure that the investments themselves do not gradually become worthless. This means that the second aspect of the "dialogue" might involve some financing by the Arab states of those very kinds of energy which will gradually come to replace, or at least supplement, oil.

Now this demands a very close co-operation in many areas. It is not enough simply to say: "Give us the money, and we will try to see what can be done next – with a guaranteed share of the profits for you of course". The investors themselves (in this case, the Arabs) will want to know which sources are likely to prove the most fruitful over what kind of period; how far they can, and how far they cannot, replace oil; what kind of return they will have when; how far *they* can share in the advanced technology that might go with it. In other words, they might want to be in on the basic decisions from the very beginning. And if these decisions have a particular regional bias – for instance in developing the resources of the whole Mediterranean area by a combination of Western technology and Arab money – and if everyone concerned wishes to ensure that the Mediterranean does not die of pollution in the process, it means that there has to be not merely a dialogue but an intimate dialogue. It also means involving all the countries with Mediterranean sea-shores in the process – even including Israel. So the consequence is that the "Euro-Arab dialogue" involves very close collaboration between the Europeans and the Arab states; it also means that they *all* have an interest, in the not very long term, in peace in the Middle East.

This does not mean that there are no problems. It is in fact a very difficult path. But it does seem to me to be greatly preferable to the alternative, that of a kind of semi-permanent confrontation between the interests of the oil exporters and the oil consumers, which would in the end benefit neither, nor even create the minimum interest in a peace settlement between Israel and its neighbours which might exist at present. But it also means something else.

One further interest of the OPEC countries has been in the development of the "downstream" operations which have up to now been the preserve of the oil companies. The interest that the OPEC countries have is frankly in the profits to be earned. But in so far as OPEC has come to resemble a cartel – an organisation of many companies which have agreed to sink their initial competition in the higher interest of maintaining prices and profits – it is going to be like other cartels: it is going to want to diversify. This means that it is not going to put all its eggs in one basket, but will try to enter as many fields as possible, provided that they are related to each other, which help it to make money. This, after all, is what the oil companies did. As OPEC begins to do the same – and it is beginning to – it will try to engage in the transport, refining and distribution of oil. When it does, it will also become involved in a still closer dialogue with its customers about the uses of oil. "What kind of oil do you need? Where can we refine it? In what way do you propose to balance the costs of transport and refining, say, a heavy and distant crude oil against the profits from the products you expect at the end of it? What about the limitations imposed by anti-pollution laws." In all these respects, the OPEC countries will talk directly to the consuming countries, which is a direct threat to the interests of the oil companies. It is they, up to now, who have taken this kind of decision. From now on, they will find themselves in direct competition with the oil states; and the oil states have the immense advantage that they can fix both the prices and the production of oil to suit themselves. It is likely therefore, especially in Western Europe, that the interests of the oil-exporting states and the oil-importing states will come to correspond more and more closely; and the old identity of interest between the oil-importing state and the oil company will gradually be broken. This identity of interest was in any case in itself an outgrowth of the pattern of imperialist domination against which the Middle Eastern countries in particular have so recently and successfully revolted. It has not been broken yet. The oil companies are still highly successful and highly profitable intermediaries between producers and consumers. But their profits, now made largely at the expense of the consumers, do make them more vulnerable, as the OPEC countries engage in the downstream business for themselves. As I suggested earlier, the oil companies, through their flexibility and their political expertise, do provide a valuable function in screening the oil business from other kinds of politics. But if real progress is made towards peace in the

Middle East, the value of this function will decline, and its liabilities will increase.

The kind of interaction of interests which is involved here raises new questions about an old notion. I have already mentioned that the initial American reaction to the events at the end of 1973 and the beginning of 1974 was to propose an agreement on joint policy between the oil-consuming states. The European countries were on the whole unenthusiastic – and the reasons are now clearer. It was not only a question of avoiding an immediate confrontation with countries on which they depended for their oil; it also implied a recognition of the need for a long-term dialogue over a wide range of issues in which the interests of the Europeans and of the Arabs could move towards a system of mutual support. In so far, therefore, as any proposal for "joint action" by the oil-consuming countries is going to imply an active form of hostility to the Middle Eastern states, it is unlikely to be acceptable. But at the same time, the oil-consuming states need to make themselves less vulnerable to a sudden boycott or to the pressure which might arise from the threat of one. To this end, they *have* agreed with the United States that in a future emergency, there should be a common policy by which countries who depended on imported oil should help each other out by moving their supplies around, and making sure that no single one of them was singled out for the starvation which might make it succumb to political pressure from the oil-exporting states – especially those in the Middle East. (France has refused to join in this venture, but the other principal countries of the industrialised West all support it. For once, French opposition is not likely to make very much difference.) This is a sensible step. It means not only that if a boycott is less likely to work it is less likely to be imposed – though that is important in itself. It also means that it should be possible for the Western European countries to *combine* the approach indicated in the Euro-Arab dialogue with that of being prepared to resist undue pressure if need be. Both psychologically and politically, it is a healthy combination. It is of course vulnerable – above all to the kind of pressure for a boycott which could arise if new hostilities broke out between Israelis and Arabs. Indeed, such a boycott in such a context might mean the end of the Euro-Arab understanding which is now being so laboriously built up, and could lead instead to an outright and very dangerous confrontation. But short of that, there are grounds for modest hope.

This hopefulness might be even stronger if the nations of the Western world were able to come to an agreement with the Arab states on the use of the dollars now accumulating in vast sums from the sale of oil. As I have suggested, these dollars are not being removed, either from Europe or the United States. But they are not all being put to use either. The problem here is that of establishing a framework for international agreement. The proposals that have come out of the United States so far seem to envisage a vast international bank to operate these funds, but one that would effectively be under Western control. Not surprisingly, the Arab countries are not interested. On the other hand, it is clearly important, both for the Western economies, and for a future political understanding between the Arabs and the West, that these petro-dollars as they are called, are not moved around from currency to currency, short-term investment to short-term investment at the whim of the ministers in tiny states, or through politically motivated malice. Each side, therefore, needs some degree of control – and especially if the Euro-Arab understanding is to bear any fruit in the paramount question of how to help the developing countries of the world to avoid the starvation which now threatens them. So far, very little progress has been made on this vital issue. The muddle which persists late in 1974 merely suggests that the United States and the European countries are becoming more vulnerable to another form of pressure. Not only do they have to face the possibility of an oil boycott, but they also have to reckon with the fact that their system of money can be wrecked by the financial ministers of the Gulf. The fact that this pressure is very distinctly *not* being applied may be one of the most hopeful signs yet that the Arab countries are more interested in co-operation than in conflict. It might even be the case that their interests are reciprocated.

For what has happened in the end is this. The success of the general revision of the politics of oil, which has affected the money system of oil, and through that the entire economies of the Western world, began as a political revolt by the OPEC countries against the prevailing assumptions on which the oil industry operated. But its success has now brought about a much greater necessity for political understanding between the producing states and some consuming states. This new understanding might affect the structure of the oil industry profoundly; it might also create a wider range of common enterprise between the Western European and the Arab states than has existed for many hundreds of years.

It is not an unhopeful prospect – except perhaps for the oil companies. It will certainly mean a series of painful political and economic re-adjustments in Western Europe. But it can also mean much closer collaboration with their neighbours in the world – provided that there is a real chance of peace in the Middle East.

But what it does not mean is that here is an end to the problems of energy. For the basis of potential understanding is still that oil now costs more than most countries can afford for very long. The search for other fields of oil or gas is on in earnest in the world; so is the search for other forms of energy. For this reason, the oil-exporting states know that they have very little time in which to create a pattern of co-operation with those who once dominated them. They are trying to make the most of the time available. But in doing so, they are inevitably stimulating the search still further. It comes back to the balance of tensions between short-term and long-term considerations – a balance which operates on both sides. What happens next will be determined by the success of the search for new sources, and perhaps new forms, of energy.

6

WHAT HAPPENS NEXT

The world is now in a very peculiar situation. Through many years of growing reliance on oil, it has reached a position where the economic survival of many countries, and perhaps the physical survival of millions of people, will depend on whether enough oil is available or not. At the same time, it is clear that there is enough oil in the ground to meet all the world's needs for the near future. Yet the location of this oil, and the political difficulties that surround it, make it an open question whether it will go on being available to many countries – at least at any price that they can even hope to pay.

The countries which are hardest-hit by this position, and have already suffered enormously from the price rises of recent years, are those in the poor areas of the developing world. The anxieties of the Western Europeans, or the Japanese, over their balance of payments are well justified, and they do give grounds for fear that a world-wide economic slump might be coming on fast. But these anxieties are nothing compared to the agonising questions faced by governments like that of India or Bangladesh, which are trying to ensure the survival of today's infants by a food programme which depends on oil products that they simply cannot pay for. But it is still one of the facts of life that questions like this, even though they are of the greatest urgency, will be resolved – if they are – not on their own merit, but as a result of other decisions. These other decisions break down into three categories. The first category is essentially political. Can the advanced countries of the world, especially those of Western Europe and North America, reach a reasonable understanding with the oil states, which will make sure that the actual physical supply of oil continues? (There is also a secondary question involved here, which is whether the Soviet Union would try to queer the pitch, but I shall discuss that later.) Here, I have suggested that the outlook is not unhopeful; at the same time, even if a policy of co-operation does survive the political difficulties which exist, it will not in itself change the price of oil.

That suggests the second category. If the price remains as high, or nearly as high, as it is at present, there is going to be a growing need to rescue the world economy from the threat of total disruption or a full-scale slump. Only if this is successful can the developing countries expect any relief from their catastrophic situation. Part of the solution will lie in the forms of co-operation between the oil states – especially those of the Middle East – and the Western world. If the Middle Eastern countries do agree to continue re-investing a large proportion of their huge oil revenues in the West, they will be able to avert disaster. But this is only the first step. After that, the real task will be to channel money into the developing countries so that they can afford to buy oil and its products once more. In the end, this has to be a decision which only the Arabs can take. So far, as I suggested earlier in this book, they have not shown much concern – though they have begun to discuss some forms of selective help in the name of Muslim solidarity. But to encourage a more general and thorough-going form of aid, it might be necessary for the Western countries to show that if they themselves are going to be reasonably sure of maintaining their present very high standard of living, they are also prepared to be more generous to the poor countries too. A *concerted programme* of Western and Arab aid will become a necessity.

And this concerted programme means aid in money in the first place – largely from Arab sources – but also aid in economic and technical terms, which at present only the Western world and the Soviet Union are qualified to give. It is in this sense that the decisions on which the hopes and lives of millions in the world's poorer countries will depend have to be reached first among the richer nations. It is perhaps also true in a wider sense that, the way the world works at present, the best hope for the poor is that the rich stay rich. The economic development of the poor depends very heavily on the ability of the rich to go on stimulating world trade, to inject fresh economic force into those socieites which in general are still living on the land, and to generate more technical help. In this sense, too, those whose needs are most urgent have to wait for agreement among the others.

The third category of decision centres on a question which has by now become familiar in every newspaper. Can new sources of oil, or new sources of energy be found? If so, might this make oil cheaper? Could it also help to *break the political connection* which now exists between the question of reserves and the question of prices? Alternative fields of oil, or new kinds of energy, could help to create a situation in which there was a much more general distribution of oil reserves, and so bring prices down by simply having more oil available (economic "laws" at work, undistorted by politics) and by giving the big oil-importers more sources to turn to. In which case more oil could be released at a more reasonable price for those who can scarcely afford it at present.

What I have suggested so far is that the most urgent problems created by the present oil situation are faced by the developing nations; but they can only deal with these problems when the industrialised societies have already begun to solve theirs. For this reason, the bulk of this chapter will be concerned with the kind of action that can be taken in the West. It runs the risk of appearing to overlook the most acute and miserable questions of all but the priorities here concern possibilities and not problems.

The last category – that of discovering new sources of oil or new kinds of energy is the one I shall discuss first, if only because that is the way that most people have by now become accustomed to think about the "oil crisis" of the past few years. The assumption generally is that things will remain pretty impossible until these new factors come into play, and that thereafter they will return to something like normal. This is a view which is perhaps most frequently expressed in Britain, where the eternal promise of the North Sea oil has led people to expect that after about 1980 Britain will be rich again, and might even join OPEC (though there is a pretty slim chance of that, for the reason I mentioned in the last chapter). There is even sometimes an ill-concealed feeling that then the British can really tell the Arabs where they get off. But this kind of attitude is not confined to Britain. In the United States, under the impact of the warnings of an "oil crisis" and then of a more general "energy crisis" the American government began to spend vast sums of money, both on opening up new oil fields and on researching into new forms of energy. The objective here is that the United States should be more nearly self-sufficient in its energy requirements by the Year 2000, and that in between it should not become so vulnerable to anyone's demands as to have to revise its foreign policy seriously. And in Western Europe more generally, there is also the widespread hope that North Sea oil – of which the Norwegians have an enormous share, and the Germans and the Dutch considerable proportions – will be enough to reduce dependence on the Middle East very sharply.

How well-founded are these expectations? To consider this, it is worth distinguishing between the two kinds of development that are going on: new sources of oil, and new kinds of energy.

OIL

The first, and most fundamental point to note about new oil fields is that they are going to be very expensive. The second is that they will be under the national control of some of the countries that need them the most, or at least under the control of their friends and allies. It is in fact only this aspect of friendly political control that makes it worthwhile to develop them at all. For, in discussing their value, one should never lose sight of their real cost as opposed to the basic cost of the oil that can be raised from the reserves that already exist. The real cost of oil in Saudi Arabia is ten *cents* a barrel. So far, it has cost a million *pounds* a sample to bring North Sea oil ashore in Britain. Obviously, this kind of comparison does not mean much by itself since the development of *any* oil-field anywhere is vastly expensive in the first place. But that is precisely the point. Governments like those of the United States and Britain – or other countries in Western Europe, or Japan, which is heavily engaged in the search for oil in the South China Sea – are prepared to spend enormous sums in order to secure their own access to oil, rather than rely on the plentiful reserves which exist already. The North Sea field, the Alaskan field, and the potential fields which exist elsewhere, all demand tremendous investment and whole new technologies before they can be exploited. In the case of the Alaskan field, this has meant in addition some considerable political strain in the relations between the United States and Canada – since the most economic way of bringing Alaskan oil to the United States cuts across Canadian territory, or through Canadian territorial waters. Here, the American government has not always shown itself much more sensitive to the national sovereignty of its friend and neighbour than the companies did in the past to the Arab states.

Now, if governments are prepared to undergo the great expense, meet the technological demands, and encounter the incidental political difficulties which are involved in opening new oil-fields, they are obviously running a particular risk. Suppose the OPEC countries, instead of suddenly raising the price of oil as they have done in the past, were suddenly to reduce it to something like its real cost? If they saw that their markets were threatened, and their chief source of revenue disappearing, this would be the obvious thing to do. The oil from the new fields could then never compete against the cheap floods that were available from the Middle East; and the fields themselves might be seen to have been such a gigantic waste of money that they would reduce the Concorde to the proportions of a pensioner's modest little flutter on Derby Day. And indeed, even if the process were not so sudden, but the much more likely one of a gradual decrease in the price of oil from the existing reserves as the new oil-fields came into operation, what would the outcome be? In a sense, the new fields would have served their purpose: they *would* have brought about a reduction in the price of oil, and everybody, especially the poor countries, would benefit. But in another sense, new complications would ensue.

For the exploitation of the new oil-fields has to be made commercially attractive. That is to say that somebody must expect a reasonable profit from them if the money is going to be put up in the first place. It is no use expecting the governments to do it all as a kind of gigantic national enterprise: they simply don't have that kind of money; and if they proposed to raise it by heavy taxation, they would soon be voted out of office. Or else they would have to cut down so heavily on other services that they would cease to govern effectively. So the money has to be raised privately: from the old oil companies or new partnerships, which means in turn that a near certainty of a handsome profit is a pre-condition to the search for new oil. Now what this means is that all those governments concerned in the search for new oil *are now acquiring an interest in very high oil prices.* Because only if prices remain high can the oil which the governments hope to produce stand any chance of competing with Middle Eastern oil. If, on the other hand, prices do remain high, their own oil can sell – and sell at a sufficient level for profits to be made.

This is a very curious situation. A government like that of Britain is borrowing itself into near-bankruptcy in order to go on importing oil. A drop in oil prices would be the best thing that could happen to the British economy. But later on, if oil prices did drop, Britain's import bill would be much less affected, since it would by then be producing most of its own oil, while on the other hand, all those parties (including the government) now engaged in exploiting the North Sea, would find that they were losing a heavy proportion of what they had invested. Indeed, to the extent that Britain exported oil, a drop in prices would mean that the country as a whole lost money. The hope, therefore, is that Britain will somehow muddle through again until it produces most of its own oil requirements, but that by then prices would still be high!

But the situation is in fact even more curious than that. For, long before the new oil really begins to flow, and before prices begin to drop (if they do) it is now a major interest of those engaged in exploiting the new fields to make sure that Middle Eastern, or other imported oil, does not undercut domestic prices. This is most noticeable in the United States, where the policy of the American government has been to keep the difference between prices for American oil and prices for imported oil to a minimum. That means, as it meant before in a similar situation that I discussed earlier in this book, that Americans are being asked to pay more for the oil they use than they would otherwise have to. Imported oil *has* to be expensive if it is going to be worthwhile developing the Alaskan and other oil-fields at all. The alternative, of course, is to restrict the amount of oil that the country is allowed to import anyway – the quota system that was first imposed in 1959. But the needs of the United States for imported oil are now so great that quotas are harmful to the economy, and they were abolished in 1971. But, if they have gone, price control – a control to ensure that prices stay high, not that they come down – is the only way of making sure that enough money can be found to develop America's own resources. Price controls of this nature work of course to the advantage of the oil companies, and is one reason for the vast profits they have been making in recent years; but, quite apart from the power of the companies, they are also necessary, unless the whole system of raising capital in the Western world is to be changed in a revolutionary manner.

From these examples, one can now perhaps see where the attempt to bring the price of oil down by making more reserves available has in fact begun to lead. It is leading to a system of keeping the price up! And it is necessary to emphasise again that this is an outcome of the political considerations involved, not the economic ones. It is the *political* desire to be as independent as possible of the Middle Eastern oil states which provides the chief motive for developing new oil fields; and it is probably foolish to expect that this political decision could, by itself, do much to reduce the cost of oil in the world today. It might even lead to a political conflict of the opposite nature to that which characterised the oil crisis of the last few years.

It might in future be to the interest of the Arab and other states to reduce oil prices; and to the interest of the Western states to keep them high. If such a conflict did become earnest, the only way in which the Western world could hope to win would be to reintroduce a quota system on imported oil, so as to limit very strictly the amount of Middle Eastern oil that was used, and make sure that the Western product was not undercut in the Western market. But this would then create new political risks and complications: the risks of further conflict with the Middle Eastern countries, which few people can seriously want, and the complications of having to cut down on consumption of oil at home.

For it is important to remember that no-one is going to be self-sufficient in oil for the foreseeable future. Even in the case of the United States all the expectations are the other way round. The country is expected to go on increasing its demands, even if the economy doesn't grow very fast; and even if demand does not grow as fast as it has done in recent years (if it did, the USA, the country with the world's biggest oil industry would also be the world's biggest oil importer!) it would still mean that the country needed Middle Eastern oil. As for other countries like Britain; there are several matters to take into account before one talks too glibly of being self-sufficient in oil. First, the oil fields of the North Sea are of course large and the estimates of their size are growing. But this does not mean that they will satisfy all Britain's oil requirements. By even a modest estimate, Britain will produce as much oil as Kuwait in less than ten years' time. But the oil is light, and will not meet all the country's needs, especially in industry. Britain, even if it ends up as a net exporter (a country exporting more oil than it imports) will still have to import oil. And by far the most likely area to import oil from is the Middle East. Even if the most optimistic estimates of the North Sea's potential turn out to be correct, British industry would be unable to function unless oil were still imported. And this still means a degree of dependence on the Middle East – the best way of dealing with which would be to try to turn it into a form of honourable interdependence, rather than nurse the hope that one day Britain will be able to ignore the Arabs and that they will all go away. Similar considerations apply to other countries.

So, if no-one in the Western world is going to be really self-sufficient in any foreseeable future, the risk of a long drawn-out political conflict with the Middle Eastern states is still unacceptable. In consequence, the question of oil prices will need to be solved by international consultation and co-operation rather than by a competitive taking of sides. Quotas, price controls and all the rest are short-term, unsatisfactory and potentially dangerous instruments of national policy and they would clearly do nothing to alleviate the problems of the developing world, which, in any larger view, is also in the national interest of the advanced countries to do. What one might have to accept is that prices will come down, and the extent that they do so will represent a loss on the investment that is now being put into developing the new oil fields. But none the less, the new oil fields will then still represent an overall gain: they will make oil cheaper, both in the West and in the developing world. It is these overall terms, not in terms of a fictitious independence that they should be thought of – and still less should any dreams of independence be accompanied by fantasies of cheap oil. The more independent, the more expensive the oil. The cheaper the oil, the more it must represent a continuing form of co-operation with the world's other oil producers, especially those in the Middle East.

These are some of the implications of developing the new oil fields. They represent a very major hope, both for the industrialised countries of the West in relieving them of the tremendous strain their balance of payments has to face at the moment, and for the developing countries in making oil cheaper again. But they do not relieve anyone of the need to find a way of co-operating with the oil-producing countries of the present day. In other words, the political requirements remain, and I shall return to these later in this chapter.

OTHER FORMS OF ENERGY

But the search for new fields is only one part of the immense activity that is going on now, especially in the United States, in the attempt to reduce reliance on imported oil. There is a much more general attempt to find new sources , or new kinds of energy altogether.

This search raises different kinds of questions. It will not imply – in the first instance anyway – a set of decisions about prices of oil, though here too, it should be borne in mind that many of the attempts that are being made to utilise other kinds of energy are worthwhile only if the oil prices remain high. But the basic questions relate to two other matters: first, what is the general cost; second, how soon will these forms of energy become practically available?

Here, one should break the search down into two distinct categories. The first is for forms of energy that are really very like oil, but have not up to now been thought worthwhile to invest in. Indeed, in this first category, one source *is* oil, but oil that is not found in liquid form under the ground whence it can easily be pumped up, but oil present in shale rocks on the surface, from which the process of extraction is much more difficult. It is also very much more expensive. All the same, there are very large areas of oil shale in the United States and the process of exploitation has begun, because, once more, it is part of an attempt to become more self-sufficient rather than to find a cheaper energy.

A second, somewhat similar, source is that of the tar-sand deposits, again found in great quantities in the United States and Canada. Instead of taking the oil out of rock, the problem here is to take a heavily dispersed form of carbon out of sand. It is useful for some purposes; it can be turned easily into heat, or used to synthesise some oil products. But in the case of tar-sand, like oil-shale, the basic question is one of self-reliance in political terms rather than cost, in economic terms.

Beyond those two forms of exploitation, which would have been thought eccentric even a few years ago, there are of course the much larger possibilities of natural gas and coal. But in most countries, as I suggested at the beginning, coal is by now so much harder and more expensive to get at than oil, that governments do not wish to re-invest heavily in coal-mining if they can help it. Coal *can* be used to synthesise petroleum and oil products over a fair range of uses, but it represents a very poor return on the capital invested. And the question of how much coal a country actually has in reserve after the general exploitation of coal in the world over the last one hundred and fifty years, is one which varies widely. In Japan, for instance, it is simply not worthwhile trying to go back to coal; in Britain, and in the great coalfield which extends through parts of Germany, France, Luxembourg and Belgium, the events of the past few years have certainly reprieved some mines and helped to promote a degree of modernisation in others, which would not have been worthwhile if oil prices had not risen so steeply. But in general, nobody is going back to coal as a prime energy source, and if everybody did, there wouldn't be enough coal anyway. There are two partial exceptions to this: the United States and the Soviet Union.

Both have very considerable coal reserves, and both could, if they chose, allow much more of the economy to depend on coal than at present. But the pattern of dependence on oil which has been growing in both – in the USA for generations, in the Soviet Union very rapidly since the Second World War – shows no sign of being reversed. And in the United States there is a particular problem too. The cheap coal that is accessible there is largely near the surface of the earth, in long horizontal deposits. That is, it can be mined, and can only be mined cheaply, if the surface is torn up in huge sweeping operations from which it is practically impossible for the surface-earth to recover. This method of strip-mining has devastated large parts of West Virginia and Tennessee; and no-one who has seen the results would want to see more of any country turned into the kind of wasteland left there. (Especially when at least much of the land could be used for farming instead, and when even today, tenant farmers are literally bulldozed off their land by the companies that own the mining rights.)

Natural gas is a different proposition, being cheap and practically ready to use. Gas from the North Sea already does much of the cooking and heating in countries like Holland and Britain. But here too, the question arises of how much there is. In the United States for instance, the search for natural gas deposits has been intensive; but, by now, so much is being used that for the past three years, the rate of consumption has actually been greater than the rate of finding more. The American government has in fact begun to import considerable quantities from Algeria. And as for the Soviet Union, even with its own huge deposits of natural gas, it has also found it convenient to import some from Iran. France likewise imports from Algeria. In other words, natural gas is an energy form which exists in much the same political context as oil; and though it is a very valuable alternative in itself, it should not be thought of as something which can do much to change the oil situation in either the short or the long term.

These different energy sources represent the first category: those which resemble oil in the sense that they have to be developed in a similar manner or in a similar context. Most are merely more expensive ways of doing much the same thing as can now be done more cheaply by oil. In the case of all but natural gas, there would be little interest in them, were it not for the fact of the recent very abrupt rise in oil prices, and the consequent desire on the part of many governments to achieve some independence from the oil producers. In the case of natural gas, on the other hand, the quantities and distribution of the reserves combine to ensure that the political context is much the same as that of oil anyway. All in all, one might argue that these forms of energy do not really offer an alternative to reliance on oil so much as they represent a reaction to the questions raised by the politics of oil in the first place. And if that is so, the questions of how cheap? how soon? answer themselves: these resources can be called upon fairly quickly, but at very great cost. (Again, apart from the different question of natural gas.) They too suggest that it is the political relations between the oil producing countries and the oil consuming countries which are likely to be of the first importance in the future.

But there are also longer-term kinds of energy resource which might provide a genuine alternative to oil. This is the second category, but here the questions of cost and time scale become even more important. The most generally popular of these is nuclear power.

As a term, nuclear power itself covers two quite different processes. One relies on splitting atoms – nuclear fission – and the other relies on welding different atoms together, and in the process releasing energy from their spare components, so to speak. That is nuclear fusion. Nuclear fusion is perhaps the answer to the long-term needs of humanity. It is the process that powers the sun, and if it can ever be mastered there would be almost limitless supplies of energy available from the oceans. But it might just as well be forgotten about as a workable proposition for the rest of this century. The difficulty is not in bringing fusion about – that has been done many times in the fireball of the hydrogen bomb – but in controlling it – as the hydrogen bomb itself makes clear. To control it would mean to maintain a temperature of between 100 million and 1,000 million degrees Fahrenheit for up to two hundredths of a second – which in terms of the time of nuclear particles is about the equivalent of an ice age. No-one is anywhere near this advanced stage of technology yet, and though two or three different technological approaches are being explored, there is no prospect whatsoever that they will bear any fruit until well into the twenty-first century.

The other form of nuclear power is that of fission. This has of course been used extensively since 1945. Indeed, Britain has been the world's leading consumer of electricity based on nuclear fission throughout the entire period, though it has just been overtaken in absolute terms by the United States. There is no doubt, in fact, that fission-generated electricity will expand everywhere in the next few years. *But* it is a very expensive process. In terms of the economy, it represents one of Britain's many major disappointments in the years since the Second World War; and the basic difficulty is the same everywhere: nuclear reactors use up all the fuel they are given. The production of this fuel is not cheap, and unless one can produce reactors that actually produce more fuel than they use, fission power will continue to be very expensive. But the new reactors which produce more than they use, the breeder reactors, are everywhere the subject of experiment: in the United States, Western Europe, the Soviet Union, Japan. At present, it looks as if the Western European experiments might be more fruitful than the others. But here, too, one should bear in mind that even if everything goes well, there can be no widespread dependence on breeder power until towards the end of this century. By then, it could provide a fair proportion of electricity needs, at least in the more developed countries. (High time, too.

The world's resources of natural fuels might be running fairly low by then – though this is a point to come back to.) But, even so, there will be limits on the extent to which this form of energy can be used.

The fact is that fission reactors have very serious drawbacks. In the first place they generate enormous quantities of heat. So far as the present technology is concerned, or any technology that is even being discussed, this heat will be drawn off through water. Hot water cannot be dumped indefinitely into rivers and seas: the fish die – or at least stop breeding. When that happens, other forms of life start flourishing which can ultimately prove gravely damaging to both the water and the air around it. Indeed in an algae-bacterial form, they could even threaten the survival of large cities near rivers or lakes. So it is not *merely* a question of respect for fish and wildlife but one of human survival too. This is the first limit imposed upon the use of nuclear reactors.

The second is of course that as a by-product reactors give off a lot of radio-active waste. This can remain active for hundreds of years – in some cases, even thousands. All kinds of precautions are taken in the disposal of such material: it is sealed in the thickest of concrete containers and sunk deep at sea, or disposed of in other equally cumbrous ways. But wherever they are dumped, there is always the danger that a container will burst (the energy in a small underwater earthquake would be quite sufficient to damage the thickest concrete), or the precautions found faulty. And then there could be a local catastrophe, or a widespread one, depending on whether the radioactivity gets into the atmosphere or into the major ocean currents. The danger is already there, lurking on the bottom of the ocean or buried under the earth; but the extent to which one can add to it merely because the precautions have proven effective so far must always remain open. And this too places a very heavy constraint on the degree to which one can rely on nuclear reactors for massive energy supplies in the future.

In other words, nuclear energy is costly, potentially dangerous, cr, in the case of fusion, not likely to come into use until the lifetime of our grandchildren. What other sources are there?

There is of course the sun, To trap the radiant heat of the sun, and channel it for home heating, has been proven feasible and cheap in places where they have plenty of it – like California. For much of northern Europe, the sun would appear to be ruled out, for obvious reasons. In fact, its heat can be stored, in molten salts, or for a short period in hydrogen. The difficulty here, however, is that to do so on a scale which would make any appreciable difference to energy supplies would demand the building of large solar "farms" – which means taking up a great deal of land at considerable cost. And the best places to do this are of course deserts. The United States, Australia and the Soviet Union possess deserts. The other countries which possess them are in the Middle East or Arab North Africa but one might add that it is possible, even in Britain, to use the sun's heat for domestic heating, provided that houses are completely re-designed. One prototype house in Greenwich, lived in by the person who designed it, manages to keep a comfortable temperature going simply by using the heat of the sun and that of the natural wastes from the human body. It also grows things in the process. In other words, this is a feasible notion; but it would demand such a revolution in living habits that it can be effectively ruled out for a generation or two.

A further source of energy is the heat trapped below the earth's surface. Buildings could be designed to make use of this directly; alternatively, it can be used for steam-generated electricity. In California, once more, this is happening on a small scale. But neither individual countries, nor the world as a whole, could afford to do this on a large scale. It would mean a progressive cooling of the earth's crust just below the surface, which would mean a lot of earthquakes.

Beyond these two natural, but limited, sources, experiments are being made on others. The use of tides is by now a familiar way of generating electricity, and in Brittany, for instance, an immense barrage has been built to exploit the tides for turbo-generators, both as the tides come in and as they go out. But while this is ingenious engineering, it provides a very limited supply, and even if the coast were surrounded with such structures, in an *island,* like Britain, they would provide only a fraction of the energy needed. More advanced technological proposals include the use of hot hydrogen gas to generate electricity from magnets; the collection of solar energy in space, to be beamed back to earth from satellites; underground nuclear explosions to turn water into steam. But all these are extremely expensive, even by the standards of the present search for energy; and none of them could make very much difference in the near future.

Finally, there are other, less radical, but perhaps more realistic ways of using the resources we already have. The more advanced an industrialised country is, the more rubbish it creates. Most of this rubbish is at present burned or dumped without a thought of turning it into energy. If the heat from its burning were properly applied, it could make a significant difference to, say, the domestic heating programme of a city. It would of course demand extensive, but simple, underground engineering, and this of course costs money – but nothing like the sums involved in some of the more advanced proposals. Similarly, sewage could be used both for fertilisers and for the propagation of methane gas, thus reducing some of the demand for oil in the world, and also leaving beaches and estuaries very much cleaner than they are now. And one could also begin to give more serious thought to the design of electricity generating stations. At present, these are so inefficient that the greater proportion of the energy they produce is simply dumped in the air as heat. They *can* be made more efficient (again, at a cost); or the heat, which is now regarded simply as a waste product, can be applied to further uses.

These are serious, modest, and not too costly proposals. But in no single case or combination of cases is there the remotest possibility that the savings that would be made on the demand for oil could come about in time to make any real difference to the prices and politics of oil for the next ten years. Indeed, on the whole it would be wise to assume that the world will continue to rely primarily on oil for the rest of this century; and that the search for other energy sources is a sensible long-term measure to cope with the situation that will arise when oil and other fossil fuels begin to run out, or have to be strictly rationed. When this happens – when oil, for instance, is reserved for those uses, which, unlike heat, cannot easily be supplied from different sources – then the investment in the search for other energy supplies will be seen to have been well worthwhile. But it would be a delusion to imagine that anything happening now is going to change the basic pattern of dependence on oil, or the international relations which arise from this dependence for a long time to come.

In the last chapter, I mentioned the controversy implied in the phrase "energy crisis". The controversy is simple. If the rising rate of the world's demands for energy continues, there *is* likely to be a severe shortage of traditional forms of energy within two generations. But the question remains whether this rising demand *should* continue at its present rate. It is a natural outcome of what is happening now that the energy crisis should have been an American preoccupation; for it is in the United States that the use of energy is the most profligate. A fair-sized city in the developing world could be supplied with most of its heat, light and cooking needs from the consumption of a single modern skyscraper in New York – with its immense lifts, its all-night lighting (often purely for publicity) and its lavish air-conditioning. And while the United States is easily the most extravagant country in this respect – at present consuming something like a third of the world's energy, and going on to reach half by the end of the century if it continues at the same rate – it is also obvious that most of the industrialised countries use far more energy than they need. To give a single example, the minor convenience of a pilot-light burning continuously in a gas stove means a consumption of energy that is 50% greater than in a stove without one. And social habits, like the use of private cars instead of trains whenever possible, is a grossly extravagant, and in energy terms, inefficient form of consumption. It is also a major pollutant of the atmosphere – so much so that fish are now dying in Swedish lakes because of the way in which the driving habits of the English affect the air currents over Scandinavia. What has to be considered, in talking of an energy crisis, is therefore the question of whether the advanced industrial countries have the right to go on consuming energy at the rate they do. This is perhaps an even more important question than that of the search for new supplies; and it is perhaps about time the connection was clearly seen between the extravagant use of energy in the West and the starvation and poverty elsewhere. It is the Western habit of relying on cheap energy – a habit that was built up in the days of plentiful oil at ridiculously low prices – that has now driven the price of oil up so far that countries like India or Bangladesh cannot afford fertilisers. The energy crisis is in this sense really a social crisis, and, not to put too fine a point on it, a moral crisis.

Here too, though, it will take time for the social and moral dimensions of the energy crisis to be appreciated.

The so-called "oil crisis" of recent years might prove to have been a blessing in a thin disguise if it does stimulate some new thinking. But meanwhile it is still going to be true that, in the short term at least, the welfare of the poor will depend on the wealth of the rich. It is also true, as I hope this brief survey of the search for new oil fields and for new sources of energy has shown, that the wealth of the rich states must depend on a new set of relations with the oil states, and that they cannot hope to be able to ignore them. In which case, the questions at the beginning of this chapter can now be considered further. It is on the political connection between the reserves and the prices of oil that the economic future of the world over the next years will depend.

THE POLITICS OF CO-OPERATION

I suggested at the end of the last chapter that there is some prospect of a close and genuine co-operation arising out of the "Euro-Arab dialogue". But there are still three major obstacles to such co-operation. The first lies in the difference of approach between the United States and its major allies to such a venture. The second lies in the attitude of the Soviet Union. The third lies in differences which are evident inside Western Europe itself.

The first has by now been largely overcome. The original American reaction to the crisis at the end of 1973 and the beginning of 1974 was to try to organise an alliance of oil-importing states to deal with the OPEC, and especially the OAPEC, countries as a bloc. In so far as the OPEC states *were* attempting to use oil as a weapon in their fight with Israel, this was an inevitable and justified reaction. But it was neither inevitable nor justified when it became apparent that the "weapon" was a very secondary and temporary aspect of what the Arab and other oil states were attempting. Their desire to use oil to re-organise the entire range of their political and economic relations with the West demanded a realistic and responsive reaction from at least those countries which did depend on imported oil. This reaction has gradually taken shape, and the opposition of the United States has gradually ceased. Indeed, the United States itself is now importing oil and natural gas in sufficient quantities from the Arab world for it to acquire a positive interest in encouraging co-operation, provided, of course, that the co-operation does not affect the fundamental foreign policy commitments of the country. And since this appears to be well understood by the principal oil-exporting states, there is now every reason to expect that the "Euro-Arab dialogue" can continue with American blessing. But one cautionary word still needs saying: it is that it might be a part of the weary search for a "European identity" which has inflamed and troubled the European Economic Community for so many years that the European governments will try to make their dialogue with the Arabs as *exclusive* as possible. If so, they will be asking for trouble. The United States guarantees their security as much as it guarantees that of Israel; and if a new form of European obstinacy arises out of the political disagreements of the October war, and leads to an attempt to ignore a legitimate American interest in the development of the dialogue, there is no reason why any American government should continue to pay for the dubious privilege of protecting them at the present level. But perhaps one might discern a greater readiness to co-operate with the Americans as well as the Arabs among the present European governments than among their predecessors at the time of the October war, and in this sense one need not be too pessimistic.

The second potential obstacle lies in the attitude of the Soviet Union. There are two uncertainties here. The first is that no one really knows how soon the great new natural gas and oil fields of the Soviet Union can be brought into operation – for the cost of their development is so high that it demands outside help. The Soviet approach to the question of help, from the Americans or the Japanese, or both, seems to vary with more general political considerations. Now this raises the possibility that Soviet oil and gas might become insufficient for the combined needs of the Soviet Union and Eastern Europe by about 1980; and in that case, there is a chance of heavy Soviet competition for Middle Eastern oil in a few years' time. (This is also an added reason for the development of the new oil fields, quite apart from the question of prices.) And in this context, the second uncertainty that arises is whether the competition will not also be politically motivated. In its most alarmist version, the argument goes thus: even if Soviet demands are themselves modest, might not the Soviet government try to grab as much Middle Eastern oil as possible, purely for the pleasure of depriving the West of its most precious import? To that, of course, there can be no definite answer. But several considerations might be suggested.

The first is that the Soviet Union would have to try to get a strong hold on practically all Middle Eastern oil. It is no use making one or two generous deals with one or two favoured countries, if the reaction of the rest is going to be to sell their own reserves to the West. (Again, Saudi Arabia alone has enough to keep Western demands satisfied for years ahead.) But if a Soviet government were going to try such a generalised operation, it would have to be prepared to pay gigantic sums of money – far more than it is remotely likely to be able to afford. For after all, the Arab countries are most unlikely to reduce the price of oil to a level which the Soviet government could pay, purely for the pleasure of getting on better with the Russians and worse with the United States and Western Europe and Japan.

The second consideration is that the Soviet Union could

of course try a bit of force. But once more, could it take over the entire Persian Gulf? And wouldn't it risk a world war if it did? Nothing in the record of that government suggests that it would contemplate such a risk; and without taking over the entire area, it would merely drive the surviving states into the arms of the Western powers.

The third consideration is that the Soviet government is coming increasingly to rely on American, European and Japanese help, in both money and technology, to get it out of its permanent economic difficulties. Not only would it jeopardise this help, but it would also make it practically impossible for these countries to offer it, once they were deprived of oil. In this respect a Soviet take-over bid for the Middle Eastern oil market would be like cutting off its nose to spite its face.

In general, then, unless there is a *totally* malevolent and irrational Soviet government in control by 1980 – a Hitler instead of a Brezhnev – it is very unlikely that there would be any concerted attempt to deprive the Western world of oil, or to stop a pattern of co-operation from being built up. Of course, the Soviet government will try to compete with the Western powers for diplomatic influence, and might even try to compete for some oil later on; but these are problems that can be dealt with and managed. After all, if Western Europe has no diplomacy left, it might as well give up the idea of a dialogue with the Arabs anyway.

The third obstacle, though, is one that it might take a lot of European diplomacy to overcome, and that is the competitiveness within Western Europe itself. Since there is still no common energy policy in Europe, and since the different European states rapidly established a habit of trying to get the best individual deals they could after the October war of 1973, it is still going to be very difficult for them to work out their own internal differences. The real villains here were Britain and France in the immediate aftermath of the war; but since then, it has become clear that the basic difficulty lies in a much older pattern of French diplomacy. For several years, under de Gaulle, France attempted to build up a series of special relationships with some of the Arab countries – Algeria in particular. France also concluded some very expensive long-term deals with Iran early in 1974. Now this rasies not only a question of breaking diplomatic habits – which in itself is difficult enough with the French. It also raises the practical difficulties of fitting into a common framework a whole series of arrangements and institutions which already exist, and which could not be broken without creating a great deal of political resentment among the very people that Western Europe is now anxious to co-operate with. It is above all for this reason that France is bound to remain the centre of whatever dialogue emerges. But it is also that fact which encourages one to think that ways can be found of fitting the arrangements, and creating a common policy at the same time. After all, there is no international organisation which does not have *some* special arrangements and exceptions (certainly not the Economic Community) built into it; and while these might operate to the advantage of France (as they usually do) there is no reason to jump from there to the conclusion that no organisation is possible.

So there is no reason to despair, and some reason to hope. The politics of oil have been transformed in the past few years, and the price for this has had to be paid by those countries which until recently had been having it all their own way. But the lessons of the "oil crisis" have perhaps not been lost.

The first lesson is that the economies and prosperity of the advanced countries of the Western world now depend on being able to understand a new political relationship with many countries that until recently were colonies, either in name or in effect. The second is a longer-term one: the realisation that Western countries have been great spendthrifts and wasters, and that they will have to revise their habits in the use of energy sooner or later. The third is that the real price of the crisis, not in terms of money alone, but in terms of actual starvation, is now being paid by the developing world. And the fourth is that the Western powers are now in any case vulnerable to at least some of the developing countries – which brings one back to the first lesson.

The ultimate result of realising these facts could still be a general change in those attitudes and assumptions which underlie politics, and in particular the international politics of the relations between the rich and the poor. No longer can one assume that what is good for Standard Oil is good for America, or that what is good for Royal Dutch Shell is good for Europe. No longer can one assume that the national aspirations, the hurt pride, the emotions which are as irrational and as powerful outside Europe as they are inside, can be ignored as if they were merely the emotions of poor or powerless people. Indeed, one can no longer assume that people are powerless because they are poor. It is perfectly possible that other countries – those producing copper, say, or any other commodity on which the industrial countries depend – will have taken note of the success of OPEC, and consider similar common action against other companies not unlike the oil majors.

Equally, one need no longer assume that the Arab and other oil-exporting states, and the states which import oil have no interests in common. Behind their conflicts of interest lie some powerful long-term demands for co-operation. And if this co-operation can take the form, not only of agreement on matters which affect the two sides directly, but also a concerted effort in helping those who are still poor and who need their help badly, the oil crisis might yet turn out to have been all to the good.

POSTSCRIPT

Two additional points should be made here about the nature of the major oil companies. So far I have talked as if the international majors all belong to a single, homogeneous group; and this is true in the sense that their great scale of operations and their diversity of interests set them apart from other oil companies. But it is not true in any sense which might be thought to imply that they all think alike. In fact, their long term and their short term interests often diverge and conflict. There are two main lines of division among them. The first runs between those companies which have long-standing and direct producing interests in the Middle East, and those which have not. Those which have include BP and the American consortium in Saudi Arabia, Aramco. The other group, for instance, includes Shell. And the second line of division runs between those companies which are based in the United States and those which are not.

The differences created under the first distinction show how very much the interests of the oil majors can diverge. Shell, for example, was able to suggest in the mid-1960's that the basic price paid for crude oil from the Middle East should be linked to the general cost of basic imports by the Middle Eastern countries themselves — thus anticipating an argument which OPEC was to espouse later, and in some ways even taking it further. But this found little favour with other oil majors, especially those with historic interests in the Middle East.

And the differences created by the second distinction are equally important. American-based companies have a peculiar relationship with the American government, based on the philosophy that the State Department should be primarily responsible for representing their interests elsewhere. It is also based on the fact that they are vulnerable to anti-trust legislation in the United States, and have therefore tended to seek a close working relationship with the American government. These combined effects show particularly in Saudi Arabia, where the politics of oil are very largely conducted through a tri-partite (and, to outsiders, mysterious) set of negotiations between Aramco and the Saudi and American governments.

And all these differences have an effect on the nature of profit. The oil industry as a whole takes its working yardstick not from profits as such, but from "return on capital invested." It follows that profit figures have to be judged by different criteria in different countries according to the conventions operating there. These criteria can vary not only according to such matters as accounting conventions, but also according to government requirements as for example the rate of amortisation — or, the speed at which an initial capital investment is deemed to pay for itself in subsequent returns. The result is that when Shell declares a profit in Britain it is not doing the same thing at all as when Exxon declares a profit in New Jersey. These kinds of requirements, when measured against the basic notion of return on capital invested, often mean that what appear to be inexcusably large profits are not nearly so significant, nor quite so blatant, as one might be tempted to assume at first sight. I do not mean to suggest that oil companies are suffering, but I do mean to suggest that when one reads of their profits, it is worth checking where, and on what principles, and against what return on capital invested, they have been made. Unless one does this the figures themselves don't mean very much. Perhaps the only significant profit figures over recent years are those which can be expressed in percentages of previous profits, rather than in millions of dollars as such.

The oil majors engaged in their world-wide network of business because their obvious objective was, if they were lucky, to realise a profit at each stage — production, refining, transport and distribution. This incentive was the drive behind the integration of the smaller companies which went to make them up. But, even more important, was the ability this gave them to choose between various options as to *where* to take a profit if they couldn't realise one at each stage. A loss in transport, over one sector of the business, for example, might be offset by a manipulation of where and when to purchase crude oil at the most profitable prices in another sector of the business. Obviously the companies best qualified to conduct this kind of flexible economic analysis are the majors, and this does give them very considerable advantages over their rivals. But even then they have to translate their considerations into different markets. The so-called "rich" markets might sometimes offer the poorest rate of return on capital. For example, there was no profit to be made at all in the gasoline business in Britain in the mid-1950's; according to the text books it should not have attracted any business whatever. But if you were in it already, and had amortised your investments, the business ticked over. And you knew that eventually it would become profitable. And this in turn has meant that for oil companies the scale of operation — assured supplies of oil, and assured markets for oil products — were essential to any overall calculation of profit. The majors in particular have been vitally concerned to maintain (and discover) assured sources of supply because this was the basis of a more general approach to calculating profit.

Two of the arguments which have, I hope, emerged from the previous pages of this book are that the price of oil depends, not on the economics of demand and supply but on a network of political relations between the oil-producing and oil-importing countries; and that oil companies have continued to make great profits from oil, even during the recent "crisis", at the expense of the oil importers. Since the pages were written several new developments have occurred which indicate the changes that are coming about in the relations between producers and importers: changes which indicate that a political agreement is becoming more necessary, to deal with the complexity of economic problems. Whether it is possible is still very hard to guess.

The first development indicates how paradoxical the whole question of oil has become. For it is now clear that, far from producing an oil shortage, the panic of the past few years in the West and the Arab and Iranian determination to increase revenues *from* the West, have produced a glut. Production in the spring of 1975 was running at about 20% above requirements. True to form, the oil companies have been attempting to slow down the rate of delivering oil — for example by keeping tankers at sea for weeks — in order to keep their profits up. In the United States, indeed, they are suspected of having cooked the books so that cheaper oil from some parts of the world has been invoiced as expensive oil from the Middle East. But the glut has at last been realised. Now the result of this should surely be considerable Western pressure for a cut in prices. This could come about either through a reduction in the price of oil at its source, in which case the Middle Eastern and other oil-producing states would lose money; or it could come about through a cut in the delivery prices, in which case the companies would lose money. Is either of these alternatives at all likely?

In the first instance — a cut in oil prices at source — the full nature of the paradox is revealed. For Iran, at least, has been hinting at the necessity of another price *rise* — in spite of an agreement among the OPEC states to keep prices steady for the time being. The reason suggested by Iran is that inflation is eating up the value of its oil revenues almost by the day. But the inflation which has been causing havoc in the Western monetary system started with a general rise in the prices of commodities and was then given a particularly powerful new twist by the "oil crisis". Any further rise in the price of oil might very well send inflation beyond control and create new pressures among the Western states to retaliate against the oil producers, either by sanctions, or even (if *they* then retaliated with an embargo) by force. So in this case, not only is there as yet very little prospect of a drop in prices; there is also the prospect that the economic "laws" could prove to be their own undoing, and would need to be rapidly put to rights by political agreement.

Before considering any further the possibilities of a political agreement, though, it is worth asking whether the companies could cut the prices of delivering oil. Here, the chief portent for the future is probably not that of an action by any company — though companies might come under pressure from governments to cut prices — but an attempt to bypass the companies altogether. The Arab Emirates of the Persian Gulf have decided to market and distribute their own oil in Western Europe. This will mean that they are selling cheaper oil in direct competition with the more expensive oil of the companies. If successful, it will be the first major breakthrough of the oil producers into the "downstream" operations on which they have had their eye for some time, but which have up to now been the preserve essentially of the oil companies. It will probably also mean that the companies have to reduce their prices (and profits) in order to stay in business — especially if other Arab states begin to follow the example.

That could mean some general reduction in the price of oil. But it would probably still be a marginal one. More important is whether agreement can be reached to tackle the whole range of interlinked questions — the monetary system, world inflation, oil, food and development aid for the poor of the world — which now stand at the centre of international relations. Proposals fall into two kinds. The first is that of partial measures which could help to alleviate some of the problems, and thereby make the rest easier to solve, if only bit by bit. The two principal proposals in this category, are, first, the establishment of a general fund for investing (or as the misleading jargon word has it, "recycling") a large proportion of the vast amount of petro-dollars which have accumulated in various Arab accounts; the second is the establishment of a minimum price for oil, considerably below the present level – something over $7 instead of the present $11 plus. This second is a subtle proposal in that it *could* help to bring down the current price of oil but would, on the other hand, guarantee a reasonable return on investment to those states and countries which have invested heavily, in such difficult fields as Alaska or the North Sea, when the oil finally begins to flow. Thus, if an international agreement on these lines *were* secured it would both serve to weaken the oil producers' stranglehold on the Western monetary system in the present, and ensure that in the future they could not suddenly undercut Western oil prices and thereby play havoc with the Western economies later on. Unfortunately for those in the United States who framed this proposal, the minds of policy-makers in the oil-producing countries are certainly subtle enough to grasp its implications. So there is very little chance that it would ever be adopted. As for the first proposal – the establishment of a central investment fund – I have already discussed its drawbacks; and there is still no advance on the suggestion that the fund should be essentially under Western control. Both proposals therefore are unlikely to be accepted unless there is some prospect at least of a wide-ranging *political* agreement on the interacting issues I have just suggested. In other words, of a comprehensive rather than a partial attempt to deal with the problems.

So, from what I have just said here, and from earlier in the book, it follows that there are three kinds of Western proposals for dealing with the world energy crisis. The first, suggested earlier, is that the principal oil-consuming countries, to guard against future embargoes or shortages, should establish international arrangements for pooling their resources and, at a pinch, sharing out their stocks of oil. Most of the non-Communist countries, except France, agreed to this proposal, which established the International Energy Authority, or IEA. The second proposal, emanating like the first from the United States, was for the establishment of a minimum, or "floor"-price for oil. And the third was for a central investment fund, which would concern itself with the uses to which petro-dollars should be put.

But these three proposals did not provide the complete answer. First, there was a question as to how far they would be workable without complete Western agreement — and France appeared at first to oppose nearly all the American and other European thinking on the subject. Second, Western proposals were pointless — except, to a degree, those for the IEA, unless the Arab and other oil-producing states agreed; and for a time there was no prospect of this.

But there were even more fundamental questions. First, how far could oil — or even the whole complex of energy questions — be treated separately from all the other questions involved in the relations between the rich and poor countries: questions of other raw materials, questions of food, questions of money, and so on? Second, how far could the oil-producing states agree among themselves as to whether they should stabilise prices, or how far they should try to cover themselves against the effects of Western inflation, and the consequent depreciation of their own financial assets, by raising oil prices? But, on the other hand, they also had to consider how far they would contribute to even further and more rapid inflation, and therefore to a further depreciation of their own assets, if they raised prices too far or too fast. And third, both the oil-producers and oil-consumers had to consider whether, how, and in what ways they could best help the really poor countries of the world overcome the difficulties which *they* now had to face as a result of the oil crisis.

Between 1974 and 1975 a number of developments occurred which have at least made it possible to consider all these questions together. First, the two chief Western antagonists, the United States and France, reached a measure of agreement at Martinique in December 1974 on dealing with the questions of oil and money through a dialogue with the oil-producers. Second, the oil-producing countries reached an agreement on a short-term stabilisation of prices at an OPEC conference in Vienna; and an OAPEC conference in Dubai subsequently agreed to explore further the possibilities of a long-term understanding with the Western powers.

But the first conference among oil producers, oil consumers and some representatives of the truly poor, which was held in Paris in April 1975, was little short of a disaster. Some of the Western countries, including the United States and Britain, wished only to discuss the price and production of oil; and they had allies among the more conservative countries of the Arab world. Other European countries were prepared to discuss more general questions, including those of aid and development, but largely on a regional basis comprising Western Europe and the Mediterranean. But they had few allies among the more radical countries of the oil-producing world, such as Iraq and Algeria, who wanted to link a discussion of the price and production of oil with a more general discussion of the relations between rich and poor; to negotiate the prices that the Western world should pay for other kinds of imports from the developing world; and to engage a higher proportion of the income, both of the Arabs and of the West, in development aid and more favourable trade terms for the truly poor. The conference broke up in disorder; and it is not a distortion to say that while the United States blamed the obstinacy of the Algerians for wrecking it, everybody else blamed the United States for being both short-sighted and authoritarian.

But between April and the end of 1975, nearly all the countries concerned in the failure of the first Paris conference had the occasion to think again. This was partly due to other processes that were continuing at the same time anyway. For example, both the Nine European countries of the EEC and the countries of the Commonwealth sought to provide schemes to guarantee the incomes of the world's poor states. The EEC scheme, in particular, was concerned with stabilising the prices that poor countries, exporting food or raw materials, received, by taking a four-year average of their receipts and undertaking to levy a duty in any year in which prices fell below that average, whose proceeds would go to make up the shortfall. At the same time, the OAPEC states began very considerably to increase and extend their aid — not particularly through the organs of OAPEC (or even OPEC) themselves, but by individual contributions from countries like Saudi Arabia, even to countries which were predominantly non-Muslim like India, or other starving areas like that of Bangladesh. And not only were these measures of aid necessary; the very fact that they were necessary suggested something else. A general inflation of the world's other food and raw material resources, which had seemed likely to follow upon the rise on oil prices, had not happened. Instead, the recession in the world economy had tended to keep such prices fairly steady if not actually to reduce them. In consequence the poor countries were getting poorer yet again; but at the same time, the Western countries had to know that if their economies picked up again in the near future and a new boom started, the other material prices were likely to increase once more. In other words, the hopes of Western economic recovery could not be separated from the fear of the "commodity inflation" which I mentioned at the beginning of the book.

Both the modest hopes and the dilemmas created by these developments contributed towards a new willingness to tackle the whole range of the problems involved at a further conference in Paris, in December 1975. On this occasion, oil producers and oil consumers were flanked by some representatives of the countries which had neither oil nor money and desperately needed both. And this time some progress was made. Britain and the United States still argued eloquently for a floor-price for oil. But the real interest of the conference lay in the fact that a concerted attempt was suggested to relate the needs of oil-producers and oil-consumers and to use the IEA as a means for doing so; to examine the whole conjunction of questions of food, fuel and money in the world; at least to talk about a new economic order. Even in spite of the fact that the OPEC countries agreed to raise the price of oil by a further 10 per cent, even in spite of the fact that some of their spokesmen, notably from Iran, have argued that Western inflation is the fault of the West and has nothing to do with the price of oil, this attempt to find a way of dealing with all the related problems together is still the best hope there is that the world can begin to sort out the crisis which has been upon it now for several years. The second Paris conference did, effectively, no more than most conferences do: it set up committees which were instructed to examine the problems and report back in a year. But the real point is that a beginning was made and the first broken sounds of a common language established.

The starvation of millions, the bankruptcy of a system which produced an economic miracle but can no longer sustain it, a generalised conflict between the present heirs to the different kinds of an earlier history, are all alternatives to the establishment of a dialogue in such a language. The dialogue will be difficult. At least it is urgent.

INDEX